Cyclical Analysis of Time Series: Selected Procedures and Computer Programs

GERHARD BRY
New York University
Graduate School of Business Administration

AND

CHARLOTTE BOSCHAN
National Bureau of Economic Research

Technical Paper 20

1971
NATIONAL BUREAU OF ECONOMIC RESEARCH
NEW YORK

DISTRIBUTED BY
COLUMBIA UNIVERSITY PRESS
NEW YORK AND LONDON

Library of Congress Card No. 70-123122
ISBN: 0-87014-223-2

Printed in the United States of America

Relation of the Directors to the Work and Publications of the National Bureau of Economic Research

1. The object of the National Bureau of Economic Research is to ascertain and to present to the public important economic facts and their interpretation in a scientific and impartial manner. The Board of Directors is charged with the responsibility of ensuring that the work of the National Bureau is carried on in strict conformity with this object.

2. The President of the National Bureau shall submit to the Board of Directors, or to its Executive Committee, for their formal adoption all specific proposals for research to be instituted.

3. No research report shall be published until the President shall have submitted to each member of the Board the manuscript proposed for publication, and such information as will, in his opinion and in the opinion of the author, serve to determine the suitability of the report for publication in accordance with the principles of the National Bureau. Each manuscript shall contain a summary drawing attention to the nature and treatment of the problem studied, the character of the data and their utilization in the report, and the main conclusions reached.

4. For each manuscript so submitted, a special committee of the Board shall be appointed by majority agreement of the President and Vice Presidents (or by the Executive Committee in case of inability to decide on the part of the President and Vice Presidents), consisting of three directors selected as nearly as may be one from each general division of the Board. The names of the special manuscript committee shall be stated to each Director when the manuscript is submitted to him. It shall be the duty of each member of the special manuscript committee to read the manuscript. If each member of the manuscript committee signifies his approval within thirty days of the transmittal of the manuscript, the report may be published. If at the end of that period any member of the manuscript committee withholds his approval, the President shall then notify each member of the Board, requesting approval or disapproval of publication, and thirty days additional shall be granted for this purpose. The manuscript shall then not be published unless at least a majority of the entire Board who shall have voted on the proposal within the time fixed for the receipt of votes shall have approved.

5. No manuscript may be published, though approved by each member of the special manuscript committee, until forty-five days have elapsed from the transmittal of the report in manuscript form. The interval is allowed for the receipt of any memorandum of dissent or reservation, together with a brief statement of his reasons, that any member may wish to express; and such memorandum of dissent or reservation shall be published with the manuscript if he so desires. Publication does not, however, imply that each member of the Board has read the manuscript, or that either members of the Board in general or the special committee have passed on its validity in every detail.

6. Publications of the National Bureau issued for informational purposes concerning the work of the Bureau and its staff, or issued to inform the public of activities of Bureau staff, and volumes issued as a result of various conferences involving the National Bureau shall contain a specific disclaimer noting that such publication has not passed through the normal review procedures required in this resolution. The Executive Committee of the Board is charged with review of all such publications from time to time to ensure that they do not take on the character of formal research reports of the National Bureau, requiring formal Board approval.

7. Unless otherwise determined by the Board or exempted by the terms of paragraph 6, a copy of this resolution shall be printed in each National Bureau publication.

(*Resolution adopted October 25, 1926, and revised February 6, 1933, February 24, 1941, and April 20, 1968*)

IN MEMORY OF

SOPHIE SAKOWITZ,

WHO WORKED ALL HER LIFE IN THE FIELD OF BUSINESS CYCLE ANALYSIS

CONTENTS

TABLES

CHARTS

FOREWORD

THIS STUDY IS CONCERNED with some programmed approaches to business cycle research as they are used at the National Bureau of Economic Research. It describes a programmed selection of cyclical turning points in time series, currently under development, as well as the Bureau's standard business cycle analysis and recession-recovery analysis. The analytical approaches are sketched, the statistical measures are described, and the problems typically encountered in the interpretation of these measures are discussed. Illustrative output tables are provided in the appendixes to each chapter. Descriptions of the various computer programs, their scope, available options, and limitations are available on request in the form of mimeographed supplements, which also contain technical instructions and caveats essential for the practical implementation of the programs. The programmed approaches eliminate some of the barriers that, in the past, have restricted the use of the techniques described. It is hoped that this book will promote a wider application of the analyses.

In the description of the general approaches as well as in the illustrative interpretation of the computed analytical measures, we address a somewhat less specialized audience than that to which Wesley C. Mitchell, Arthur F. Burns, and Geoffrey H. Moore directed their original expositions of business cycle measurements. This is done in the expectation that acquaintance with the computerized approach will induce analysts to use the techniques—among them, analysts who might shy away from the investment in the detailed procedural knowledge and experience necessary for conventional processing. Since many of the technical steps and even some procedural decisions are programmed, it is possible to dispense with most of the mechanics and to concentrate on the rationale of the analysis and the problems of interpreting results. A current restatement of the approaches also permits the incorporation of some thoughts developed after the publication of the early studies, and it promotes the dissemination of some of the practical wisdom that typically accrues in the process of making mistakes, be it in operation or in interpretation. In that sense, the present book is expected to contribute not only to operational facility but also to a better understanding of the techniques and their results.

This monograph is the latest of the National Bureau's efforts to make its analytical techniques and the related computer programs available to the research community. The first publication of this type was *Seasonal Adjustments by Electronic Computer Methods* by Julius Shiskin and Harry Eisenpress, originally presented in 1955 at a joint meeting of the American Statistical Association and the Econometric Society and later published by the National Bureau as Technical Paper 12. Another, *Electronic Computers and Business Indicators* by Julius Shiskin, appeared in 1957 as Occasional Paper 57. The authors of the present study presented a paper on "Applications of Electronic Computers to Business Cycle Research" before the annual meeting of the American Statistical Association in 1960. Milton Friedman's *Interpolation of Time Series by Related Series,* Technical Paper 16, 1962, which uses programmed regression analysis, should be mentioned in this connection. Also, about two dozen standard programs for economic analysis, developed or adopted at the National Bureau, are described in the mimeographed collection "Electronic Computing Memoranda."

We wish to acknowledge the sustained and generous support that our programming and data processing activities received from the International Business Machines Corporation, both in the form of grants of electronic computer time and of cash contributions. We are also grateful for assistance received from the National Science Foundation. We believe that the Bureau's computer operations provide an encouraging example of the fruitfulness of private and public support for the development of new techniques in the field of economic research.

The authors wish to express their appreciation for the personal encouragement and support they received in their general efforts, as well as in the preparation of the present study, from Geoffrey H. Moore. Only those who have received this support can fully appreciate its value. We also want to acknowledge the constructive criticism provided by other members of the staff, Philip Klein, Ilse Mintz, and Julius Shiskin. Thanks are also due to the Board of Directors Reading Committee, particularly Frank W. Fetter and Maurice W. Lee. Finally we want to record our debt to Sophie Sakowitz, who generously shared with us her rich experience in all matters relating to business cycle analysis.

Fred Howard and Virginia Meltzer edited the manuscript. Most charts were originally programmed for, and drawn by, the IBM–1130. H. Irving Forman improved the design, processed the charts for publication, and drew those charts that did not lend themselves to electronic plotting. We gratefully acknowledge their contributions.

GERHARD BRY
CHARLOTTE BOSCHAN

1

NATURE, OBJECTIVES, AND USES OF PROGRAMMED BUSINESS CYCLE ANALYSIS

NATURE AND OBJECTIVES

THE TERM BUSINESS CYCLE ANALYSIS is broad and generic, and encompasses many approaches to the understanding of business cycles, among them construction of theories or of analytical models, historical investigations, statistical inquiries, evaluation of current business conditions, and short-term forecasting.

This book only deals with certain techniques of quantitative description and summarization of cyclical fluctuations in economic time series, which have been developed and used for many years by the National Bureau of Economic Research. Specifically, it concentrates on three basic procedures and related computer programs used in the analysis of cyclical fluctuations in time series. These three procedures, turning point determination, standard business cycle analysis, and recession-recovery analysis, have been described and discussed in other publications. The present exposition (1) gives a concise description of the approaches and their rationale; (2) explains the differences between programmed and conventional procedures; (3) provides a guide for the interpretation and use of the computer output containing descriptive and analytical measures; (4) incorporates the lessons learned from sustained experience with the traditional and the computerized approaches; and (5) provides procedural guidance for the practical use of the programs.

In recent years aggregative measures of economic activity have undergone only mild fluctuations and occasional retardations of growth. Hence it is important to point out that the procedures treated in this study refer to individual activities which often continue to show distinct cyclical movements.

The National Bureau's business cycle studies have, of course, been concerned with business cycles in a broad sense, and with the interaction of the forces impinging upon general business activity. The standardized statistical description of many individual economic activities permits us to study their interrelations, to summarize their behavior, and to use their typical or atypical performance for the characterization of business cycles. The summarizing of cyclical processes and the use of regression models are examples of approaches used by the National Bureau but not discussed here. Descriptive summarization of cyclical behavior is omitted because of the wide variety of summary measures—diffusion indexes, frequency distributions of cyclical changes or turning points, indexes of amplitude adjusted indicators with similar timing characteristics, and so forth.[1] Regression analysis, on the other hand, is such a general tool that its use for cyclical analysis is only incidental and its inclusion would transcend the specified scope of the present inquiry.

TURNING POINT DETERMINATION

The determination of cyclical turning points, which is usually performed on seasonally adjusted time series, is an essential element of the National Bureau's business cycle analysis. The identification of peaks and troughs in individual economic time series permits the analytically important distinction between expansions and contractions in these series; serves as a basis for determining cyclical turns in general business conditions; and is a prerequisite for other types of analysis, including the two approaches described in this paper. The determination of cyclical turning points is the only process described here in which the programmed approach differs substantially from previously used techniques, which rely heavily on impressionistic judgments and are subject to a number of procedural constraints. By contrast, the programmed approach operates through a preliminary determination of cycles and a gradual narrowing down of neighborhoods within which

[1] Descriptions of these approaches can be found in *Business Cycle Indicators,* Geoffrey H. Moore, ed., New York, NBER, 1961, Chapters 2 and 8 (diffusion indexes); Arthur F. Burns in *New Facts on Business Cycles,* 30th Annual Report, New York, NBER, 1950 (distribution of turning points); and Wesley C. Mitchell, *What Happens During Business Cycles: A Progress Report,* New York, NBER, 1951 (distribution of cyclical changes).

turning points are selected. The process involves several weighted and unweighted moving averages of varying flexibility. In spite of the difference in approach, the program-selected turning points are close enough to the previously determined turning points that broad findings about timing characteristics of individual series are largely unaffected. The principles and problems of the determination of turning points in general, as well as the programmed approach to this determination, will be discussed in Chapter 2.

STANDARD BUSINESS CYCLE ANALYSIS

Standard business cycle analysis provides an historical description of the cyclical behavior of individual economic time series and its relation to swings in aggregate economic activity. It is predicated on the availability of such series during several complete cycles and summarizes cyclical behavior with regard to turning points and to fluctuations during expansions, contractions, and their subperiods. Cyclical behavior is described in terms of conformity of cyclical movements in the individual activity to those in general business conditions, durations and amplitudes of cycles and cycle phases, intracycle patterns, and secular changes from cycle to cycle. Since these measures are frequently based on complete cycles or at least on complete cycle phases, the standard analysis, though strong in summarizing historical behavior, is not particularly adapted to the analysis of current business conditions, where the identification of the cyclical position may be the very problem at issue. The Bureau's standard business cycle analysis could be almost completely converted to programmed procedures, the only exception being parameters that measure the conformity of fluctuations in individual time series to those in general business activity. The computerized analysis allows the speedy processing of a great many series, permits investigation of the consequences of different options, and makes the use of the method available to those who have no experience with its computational intricacies. Standard business cycle analysis and its programmed equivalent form the subject matter of Chapter 3.

RECESSION AND RECOVERY ANALYSIS

The analysis of current business conditions is the broad objective of recession and recovery analysis. This analysis measures percentage changes in economic time series from benchmarks (such as previous peaks or troughs) over given chronological spans (such as periods of three, four, or six months). Comparisons of changes during a current expansion or contraction with those during corresponding phases of several preceding cycles permit some judgment about current economic developments—their relative strength or weakness and their typical or distinctive character relative to previously experienced patterns. Judging current against prior experiences in several strategic activities may make it possible to evaluate current business cycle conditions and to pinpoint differential characteristics. The fact that this analysis can be performed on a computer assumes special significance in this era of fine adjustments and subtle changes in governmental policy mix. The approach permits the use of a large number of time series and the speedy availability of results. This approach to the analysis of business conditions is described in Chapter 4.

All three approaches are illustrated by the analysis of monthly time series of economic activities. Employees in nonagricultural establishments (referred to as nonagricultural employment in this study) and the unemployment rate are used as examples, except when problems are more effectively illustrated by other evidence. Thus, the turning point determination presented in Chapter 2 is illustrated by data on bituminous coal production. Data on nonagricultural employment and the unemployment rate provide the basis for the output tables found in the appendixes to Chapters 3 and 4.

The reader who is primarily interested in a concise summary of the basic approaches should concentrate on the first few sections of Chapters 2, 3, and 4. If his interest is centered on the modification of procedures and on the interpretation of output, he should consult the subsequent sections of each chapter. Users of the programs who have to choose among options and prepare input should ask for the mimeographed program descriptions and input instructions.

APPLICATION TO REGIONS, INDUSTRIES, AND BUSINESS ENTERPRISES

Most of the series subjected to cyclical analysis by the National Bureau are national series and were analyzed for the purpose of understanding nationwide economic events. Many of the analyzed series measure broad facets of the national economy, such as total employment, unemployment, income, production, prices, and profits. But even when the series relate to prices of single commodities, hours of work in individual states, or labor costs in a specified manufacturing industry, the analysis can still be directed toward the understanding of the business cycle as a whole—and of business conditions for the nation as a whole—during past and current fluctuations.

The methods need not be restricted in their application, however, and recent developments make wider applications more feasible. State governments, universities, trade associations, and large companies engage in ever-expanding research activities; statistical time series have become more abundant and more detailed; finally, and most important from our present point of view, electronic computers have become widely available. These changes make it appear fruitful to reevaluate the usefulness of business cycle analysis on regional, industrial, and company levels.

Business cycle analysis of regional, state, or area data can contribute to an understanding of cyclical fluctuations in general, as well as to the understanding of the role and fortunes of particular geographical areas. For the general purpose of understanding economic change, this sort of analysis may elucidate important problems, such as the differential susceptibility of various areas to cyclical swings, their differential sensitivity to government policies, the relation of growth to cyclical instability, and the possibility of constructing sensitive indicators of current and prospective economic changes. For local purposes, cyclical analysis of regional activities may help public and private agencies to establish typical relations between local and national activity, to observe past and current deviations from these relations, to anticipate impending cyclical behavior in a given region relative to that of the national economy, and to instigate remedial action where necessary. In all such cases, both historically oriented techniques and techniques designed to assist analysis of current business conditions should be used. As a matter of fact, recession and recovery analysis is already being used in some periodical publications concerned with business conditions on the state level.[2]

Knowledge of the cyclical characteristics of specific industries can also be advanced by the application of business cycle analysis to in-

[2] See, for example, *New Jersey Economic Indicators,* published jointly by the Department of Labor and Industry, State of New Jersey, and by the Bureau of Economic Research of Rutgers University. See also Gerhard Bry and Charlotte Boschan, *Economic Indicators for New Jersey,* New Jersey Department of Labor and Industry, Division of Employment Security, 1964.

dustry data. Trade associations and business enterprises concerned with the economic affairs of their industries may well apply the tools of business cycle analysis to industry output, sales, prices, and so on. Some cyclical analyses in the field of industry economics have shown interesting and promising results.

Potentially most important, but probably least explored, are the applications of business cycle analysis to the fortunes of individual business enterprises. The technical obstacles that have prevented such applications in the past are clear enough: the lack of comparable records, the scarcity of trained statisticians, and the prohibitive cost of preliminary statistical preparations (such as seasonal adjustment and smoothing). It is also true that the sales or profit experiences of certain companies, departments, and products may be more influenced by managerial policies, comparative product characteristics, specific sales efforts, or irregular factors, than by forces that produce cyclical fluctuations. Presumably, cyclical analyses would be more useful for product classes or business activities broad enough so that noncyclical product-specific factors become submerged. They may also be more useful in some industries (such as machine tools) than in others (such as food products), and more effectively applied to smoothed than to unsmoothed data. We do not presume to evaluate, generally or specifically, the usefulness of various forms of cyclical analysis applied to company activities. What should be stressed is that some of the obstacles to such analysis have largely disappeared, and that the availability of computers and programmed analytical approaches has opened the door to fruitful experimentation.

As in the case of regional articulation, the application of business cycle analysis to industries and individual companies may enrich our knowledge of cyclical processes. The whole question of homogeneity of cyclical experiences and the differential impact of aggregative change and government policies cannot be approached without analysis of regional, industrial, and company detail. And this detail can be provided and analyzed only with the help of computerized procedures.

2

PROGRAMMED SELECTION OF CYCLICAL TURNING POINTS

PRINCIPLES OF SELECTING TURNING POINTS

FOR THE CYCLICAL ANALYSIS of time series, distinction between different segments of cyclical movements is desirable. Such distinction provides a framework for orderly description. Also, behavior can be expected to differ from segment to segment, and it is hoped that this behavior is sufficiently homogeneous within segments to permit generalized description and explanation. Plausible distinctions exist between periods of cyclically high and cyclically low levels of activity or between periods of cyclical increases and declines. Combination of these two distinctions has led to various schemes of three, four, or even more phases. Characteristically, these schemes identify the neighborhoods of cyclical peaks and troughs, and partition the upswing and usually also the downswing. This leads to sequences such as recovery—prosperity—recession—depression; upswing—boom—downswing—depression; primary rise—secondary rise—boom—capital shortage—crisis—recession; or recovery—growth—contraction. In these sequences, like terms do not necessarily describe like periods. The segmentation may be determined on the basis of inflection points of fitted cyclical curves, intersection of trend and cyclical values, maximal changes in cyclical movements, attainment of prior peak levels, or by other criteria. Most of these criteria are not specific enough to yield unique statistical results. The segmentation tends to vary, for example, with the period for which cyclical curves or long-term trends are fitted, and with the choice of functions for these curves. Also, the statistical determination of fastest changes leaves much to the discretion of the investigator. Because of all these problems, it has been a widely accepted practice of the National Bureau of Economic Research to distinguish only two phases—expansions and contractions—which are delineated by cyclical turning points.[1] While the restriction to two phases reduces the statistical problem to that of determining cyclical turns—points that can be better defined and identified than most others—it does not eliminate the need for subjective decisions. This need may, however, be further reduced by the use of programmed procedures. It is the determination of specific cyclical turning points (peaks and troughs in specific time series) with which this chapter is concerned. This encompasses an exposition of the principles and problems as well as a discussion of programmed procedures.

SELECTING CYCLES

Chart 1 depicts time series of seasonally adjusted employment and unemployment, showing numerous fluctuations. The first problem is that of determining which of the fluctuations in these series should be recognized as cyclical (specific cycles). Basically, we are looking for clearly defined swings of the same order of duration as business cycles, that is, for swings that are longer than fifteen months but shorter than twelve years from trough to trough or from peak to peak. Most specific cycles identified by the National Bureau have lasted between two and seven years. We also require the amplitudes of specific cycles to be larger, on average, than those of irregular fluctuations encountered in the series.[2] In most instances, the identification of cycles in employment and unemployment is simple. The two series show well-defined swings with fairly certain highs and lows, which are indicated by X's on the chart. Even a casual examination reveals that the observed swings are rather regularly related to the expansions and contractions in general business activity, which are indicated on the time grid of the chart.

However, a number of problems may arise in conjunction with the identification and dating of specific cycles. Take, for instance, the question of whether or not a particular fluctuation in a time series should be recognized as a specific cycle. Chart 2, panel A, shows the problem in schematic form. Should the swing a–b–c be regarded as a cycle or as part of a larger expansion a–d? What criteria should

[1] The partition of expansions into recovery and growth periods is discussed in Chapter 3 (p. 71).

[2] See Arthur F. Burns and Wesley C. Mitchell, *Measuring Business Cycles,* New York, NBER, 1946, Part I, Chapter 4.

CHART 1

NONAGRICULTURAL EMPLOYMENT AND UNEMPLOYMENT RATE, 1929–65

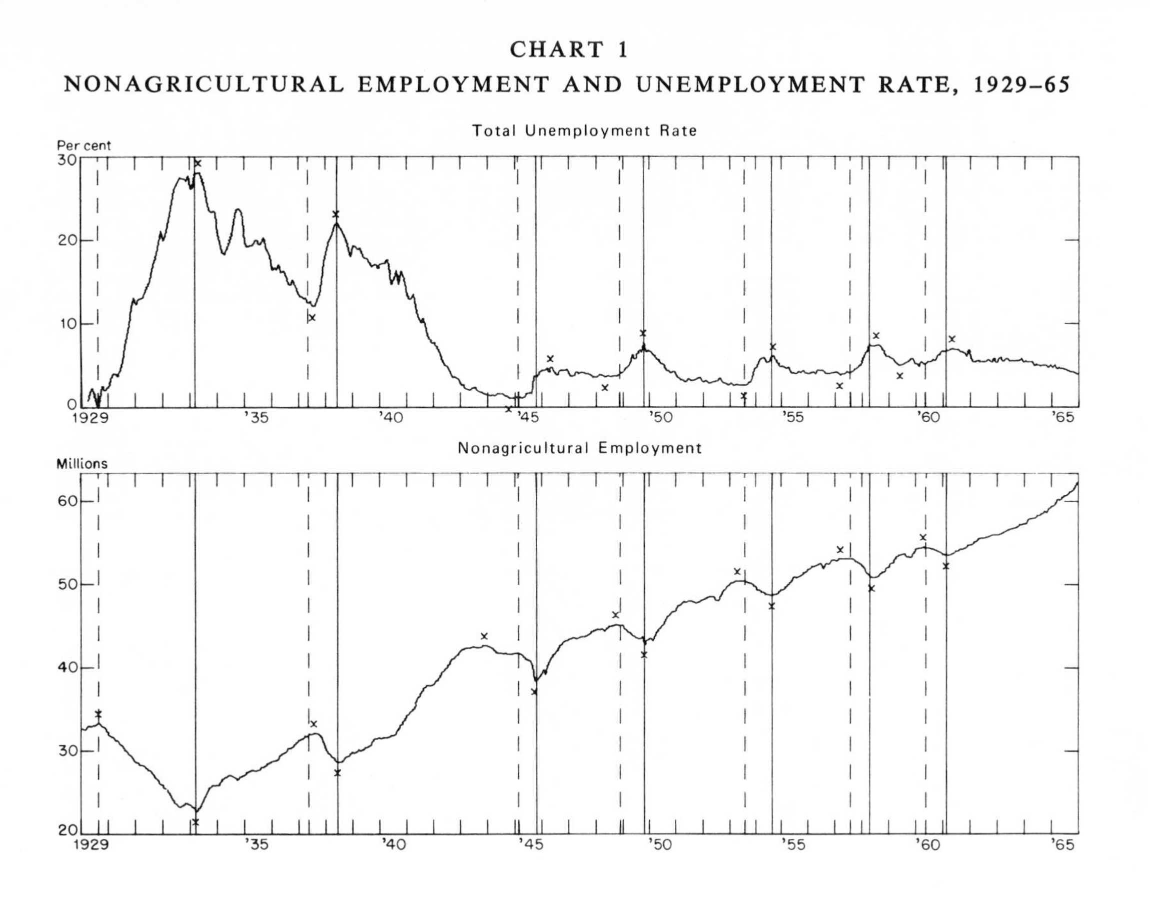

guide such a decision? One National Bureau rule is that specific cycles should have a duration of at least fifteen months. Another is that the amplitude of a doubtful expansion or contraction should not be materially smaller than that of the smallest clearly recognized cycle in the series. Chart 1 gives a practical illustration of the problem. The increase in the unemployment rate in the second half of 1959, which occurred in connection with the steel strike, appears as a fluctuation of more than random character. It is not recognized as a specific cycle since it does not approximate, in duration or amplitude, the lower limit of cyclical fluctuations in this series. A similar situation exists around 1933–34.

It is not by chance that the activities here selected, employment and unemployment, contain clear specific cycles but no example of extra cycles, that is, of specific cycles in addition to those related to business cycles. This is due to the very broad coverage of the two series, both of which reflect changes in general business activity rather than circumstances peculiar to an industry or area or activity. However, the occurrence of extra cycles is far from rare. Many sensitive series show specific cyclical declines and subsequent recoveries during the years 1951–52 in connection with the Korean War, and many activities related to the automobile industry show extra cycles during 1954–57.

Specific cycles can also be considerably longer than reference cycles. This occurs particularly when business cycle contractions are "skipped," as happens frequently in rapidly growing industries, and when the business contraction itself is mild. For a schematic illustration, see panel B of Chart 2. Specific cycles can, of course, also be unrelated or only loosely related to business cycles. This is frequently found, for instance, in series describing the harvest of agricultural crops, the exports of specialties, or fashion goods. These activities are strongly influenced by factors other than domestic business conditions.

SELECTING PEAKS AND TROUGHS

After specific cycles have been identified, it is still necessary to pinpoint specific peaks and troughs. This may raise a large number of questions, some of which have to be answered on the basis of rules which, though occasionally arbitrary, are needed in order to ensure consistency of treatment. In general, cyclical peaks and troughs are placed at the highest and lowest points of the cyclical fluctuations. Peaks and troughs alternate; i.e., a peak cannot succeed another peak

CHART 2
PROBLEMS OF TURNING POINT DETERMINATION

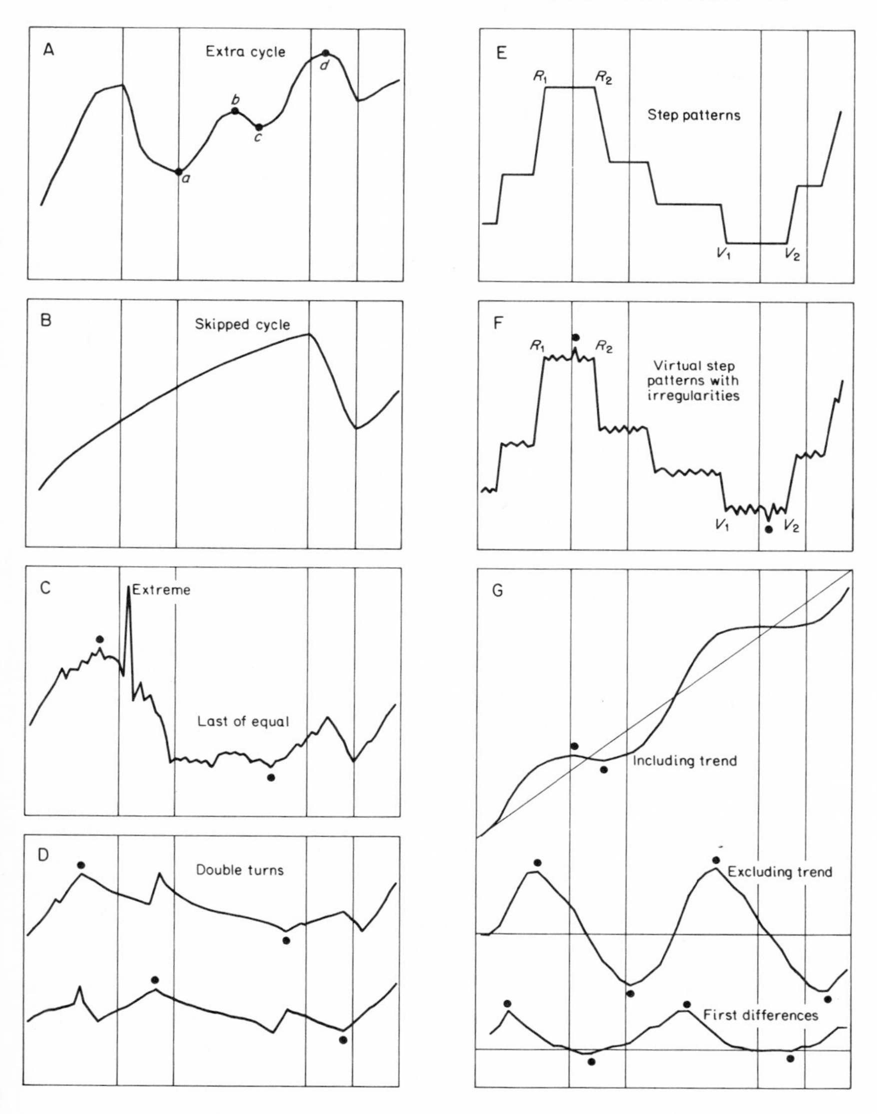

Note: Circles denote specific cycle turning points. Vertical lines stand for alternating peaks and troughs in general business activity.

without an intervening trough. Hence peaks should not be identified at the ends of series unless it is clearly possible for the next succeeding turn to be a trough; analogous considerations apply to troughs. In case of equal values the rule is to choose the last one as the cyclical turn, i.e., the month before the reversal of the cyclical process begins. Exceptions to this general rule are necessary when the values in question are clearly extreme, isolated, and possibly compensated for or surrounded by other values that deviate in the opposite direction. Panel C of Chart 2 portrays this situation and the appropriate choice of turn, indicated by a circle. On Chart 1, the unemployment low in February 1960 provides an example from historical experience. The rate in that month is lower than the lowest rate in mid-1959, but the February low is comparatively isolated, and therefore the June 1959 position is regarded as the cyclical low point. When random movements complicate the determination of specific turns, some guidance can be obtained through smoothing by moving averages. The intermediate output tables and corresponding charts of some seasonal analysis programs [3] can be of great help in deciding doubtful cases, both with regard to recognition of cycles and determination of turns. But the cycles and their turning points are eventually identified in the seasonally adjusted data, not in the smoothed series.

Sometimes a difficulty arises in cases of "double turns," that is, when a series returns to its previous peak or its previous trough level after some intermediate fluctuation. The decision in case of double peaks or double troughs is, of course, a very important one for timing analysis, since a minor difference in level and a marginal decision in the selection of turns can cause relatively large differences in timing and duration measures. The basic rules prescribe that the peak be the last high month just preceding the month in which the downward movement starts. However, if the period between the two peaks contains mainly downward movements and only one or two steep rises, the first high should be chosen. Panel D of Chart 2 depicts this situation as well as the application of the decision rule. The double turns in the unemployment rate during 1946 and 1958 do not really present a problem, since the turns to be chosen are obviously those that are later and higher.

There are cases in which, instead of showing clearly defined turns, the series maintains a peak or a trough level for several months in a

[3] Intermediate output tables of curves smoothed by a variety of moving averages can be found in the Census and BLS seasonal adjustment programs.

row. The basic rule is still to regard the last of the equal values as the turn, since the decisive change of cyclical direction manifests itself only after that month. However, if a series forms a definite step pattern in which plateaus and changes between plateau levels are common, the search for "turning points" may be inappropriate. In such instances it may be desirable to identify the beginnings and ends of ridges (R) and valleys (V), as illustrated in Chart 2, panels E and F.[4]

Some economic time series do not show actual cyclical declines, but do show clear cyclical behavior in terms of accelerations and retardations. Time series depicting economic activities with strong growth characteristics offer many examples of such behavior. The question is how such series, as illustrated by panel G of Chart 2, may be subjected to cyclical analysis. One possibility is to adjust them for trend, that is, to fit a trend line to the observations and to analyze the deviations from these trends. However, the trend will vary with the choice of the trend function, the criterion of best fit, and the time period covered. Any of these alternatives, and therefore also the incorporation of newly available information, influences the computed trend and hence the deviations and the cyclical measures. It may therefore be preferable to use a different approach and to analyze first differences or the month-to-month percentage change of the original data. When the original series undergoes cyclically regular accelerations and retardations, these derived data will show analyzable cycles.

Each solution, however, produces its own problems. First, absolute differences or rates of change are apt to show large random movements relative to the size of their cyclical component. This makes it difficult to date cyclical peaks and troughs. Second, the cyclical timing of these near derivatives differs systematically from that of the original series. First differences experience their highs at the points of the greatest absolute increase of the parent series—that is, whenever the expansion process is most rapid. The turns of these derivatives should, perhaps, be related to the points of maximum rate of expansion or contraction in the economy as a whole. Alternatively, locations corresponding to turning points in the original series could be determined by identifying shifts in the levels of first differences or rates of change.

[4] For examples of dating steps rather than turning points, see Gerhard Bry, *Wages in Germany, 1871–1945*, Princeton, N.J., Princeton University Press for NBER, 1960, p. 138; Daniel Creamer, *Behavior of Wage Rates during Business Cycles*, New York, NBER, 1950, pp. 6 ff.; Milton Friedman and Anna J. Schwartz, "Money and Business Cycles," *Review of Economics and Statistics*, Supplement, February 1963, pp. 35–37.

Such identification is simple if there are marked shifts, that is, if the original series has clear alternations of fast and slow growth. Identification of such shifts will be impossible if the growth of the underlying series changes gradually, e.g., if the cyclical component of the original series is sinusoidal rather than triangular. Economic time series are not likely to correspond to either extreme. Thus the feasibility of defining shifts in the derivatives (as approximations to turns in the original series) is an empirical rather than a theoretical question. Preliminary experiments with this approach seem promising. Further technical developments may widen the scope of its application.

Turning point determination might, finally, be influenced by the consideration of factors that lie outside the analyzed series. If one series is analyzed at a time, without reference to other activities, rigorous application of the standard rules is called for. However, in connection with a particular research project, substantive consideration may be overriding. Take, for example, the industry-by-industry analysis of the relation between peaks in hours worked, employment, and production. The steel strike at the end of 1959 affected the upper turns of many of these activities drastically. The measures of timing relations would vary in a haphazard manner if sometimes the prestrike, and sometimes the poststrike, maxima were selected as peaks. A research worker might thus be justified in basing his comparisons on, say, the poststrike peaks even if on occasion the prestrike maximum was a bit higher.

It is true, on the other hand, that such a decision might occasionally prejudice research results. For example, the arguments which suggested the selection of the poststrike peak in hours might also lead to the selection of the second peak in accession rates, although these rates typically show very early declines occurring shortly after the initial business recovery. Reasonable decisions on such matters can only be derived by an iterative process in which the growing knowledge of the subject matter is permitted to modify approaches and decisions.

PROBLEMS OF PROGRAMMED SELECTION

GENERAL CONSIDERATIONS

The importance of cyclical turning points for cycle analysis and the criteria for their selection were discussed above. Some rules were described which aimed at minimizing the role of individual judgment in the determination; yet in the formulation of these rules and still more in their implementation, individual judgment continues to play an important role. Determination of turning points can have far-reaching consequences for analysis; specifically it affects all basic measures of cyclical durations and amplitudes. Thus it is desirable to free the process as far as possible from the uncertainties of varying interpretation and from bias in the implementation of the basic rules. Progress toward greater independence from personal interpretation could be made, if it were possible to codify the relevant rules and considerations, to reduce the selection to a programmed sequence of steps, and to relegate the process to execution by electronic computer. The purpose of the efforts described in the present section is to test the feasibility of this approach.

The development of a programmed turning point determination is a process which has only recently been initiated. It may involve proliferation, tightening, or reformulation of rules, and it may necessitate some changes in the basic approach. Hence, what we have to report at this stage is provisional, much as was the case for the early programs for seasonal adjustment of economic time series. Since it is unlikely that all contingencies can be covered by any programmed approach, and since certain research objectives may require modification of rules, some overruling of the program will no doubt still be necessary in atypical situations and for special purposes. In such cases, the overruling should be explained and justified.

ALTERNATIVE APPROACHES

The technique described in this study is an adaptation of the National Bureau method. It converts this method to a sequence of relatively simple decision rules by which neighborhoods of turns and potential turning points are selected and tested for compliance with a number of constraints.

This obviously is not the only possible approach. One alternative—albeit complex and time consuming—would be the simulation of the process of turning point determination as practiced by an experienced analyst. The advantage as well as the difficulty of such simulation would lie in greater freedom when dealing with special circumstances. In such simulation, for example, turns in the neighborhood of strikes may more likely be rejected and turns in the neighborhood of business cycle turns accepted.

Another possibility would be to disregard the National Bureau method and to formulate a rigorous search-and-test procedure, based on strictly defined statistical properties of given time series. One suggested approach along these lines is based on the assumption that cyclical expansions and contractions in time series can be distinguished by the level of their first-order differences. Peaks are located where positive first differences change to negative differences, troughs where the obverse change occurs. Even cyclical changes in slopes or in rates of expansion and contraction could be similarly identified, except that here the "steps" in the first-order differences would not involve a change of sign. The statistical procedure to determine turning points and other changes in slope is a segmentation of time series on the basis of statistically significant steps in the levels of their first-order differences; the steps are selected by minimizing the variances within each segment.[5]

APPROACH EMPLOYED

The approach employed here is related to the process of turning point determination practiced by the National Bureau of Economic Research. It roughly parallels the traditional sequence of first identifying major cyclical swings, then delineating the neighborhoods of their maxima and minima, and finally narrowing the search for turning points to specific calendar dates. However, at the present time, the program neglects certain elements that are part of the traditional technique and uses some additional measures and rules.

The programmed strategy involves, first of all, the derivation of some moving averages representing trend and cycle elements only. These relatively smooth curves serve as the basis for determining the existence of expansions and contractions and for selecting the general neighborhoods of potential peaks and troughs. Local maxima and

[5] This approach was suggested by Milton Friedman, and a preliminary program was developed by Charlotte Boschan. It is particularly important in connection with the determination of cyclical phases in fast-growing series, as discussed above (see, for example, Ilse Mintz, *Dating Postwar Business Cycles: Methods and Their Application to Western Germany, 1950–67,* New York, NBER, 1970). Other approaches, incorporating perhaps some features of spectral analysis (e.g., to test the existence of cycles of a specified range of durations) are also worthy of exploration.

minima are excluded by postulating a minimum cycle duration; shorter fluctuations are eliminated in such a way that only major peaks and troughs remain. Next, the neighborhood of potential turns is redefined by identifying peaks and troughs corresponding to those of the trend-cycle curves on a time series that is only slightly smoothed by a short-term moving average. The objective here is to come closer to the eventual location by excluding the influence of values that may be several months removed from the final turns. Once the immediate neighborhood of potential turns is established on this curve, the analysis shifts to the unsmoothed data. The highest (or lowest) original values within a short span of the turns on the smoothed curve are chosen as preliminary turning points. These turns are tested for a minimum cycle duration rule and for compliance with some other minor constraints; elimination of disqualified fluctuations leads to the selection of final turns.

Consideration of this plan of attack makes it clear that numerous choices must be made in the implementation of the approach. What type of moving average, if any, should be chosen to represent trend-cycle elements of what length and using what weights? Should extreme irregular values be excluded in the derivation of these trend-cycle functions, and, if so, what are the criteria for exclusion and the rules of substitution? Within what span should a value on the trend-cycle curve be the highest (or lowest) to be recognized as establishing the neighborhood of a potential cyclical turn? Should there be minimum limits for the duration of cycles, for the duration of phases, or perhaps for expansions only? Should minimum durations be the same for expansions and contractions, and, if not, how should inversely related activities, such as unemployment, be handled? What should the values of these minima be? Should any minimum duration requirements, for full cycles or cycle phases, be applied to all derived curves, to some of them, or only to the unsmoothed data? Can amplitudes be safely ignored or must minimum amplitudes be specified?

The variety of possible answers to this sample of queries suggests that turning point determination is not simply a process of discovering "true turns." It cannot be regarded as objective in the sense that all reasonable and conscientious investigators would agree on the answers. Only agreement on the application of a specific set of detailed, and sometimes arbitrary, procedural conventions could bring about agreement on the choice of turns.

Some of the choices among procedural alternatives can be made on the basis of traditional practices, that is, by decision rules that help to maintain consistency with existing procedures. Enforcement of a fifteen-month minimum duration rule for full cycles is a case in point. In other instances, one stratagem may be given preference over another for reasons of simplicity. It would, for instance, complicate matters considerably if the programmed determination of turning points specified amplitude minima.[6] Hence, the present approach neglects amplitude considerations, except for those implicit in the various smoothing processes. Other questions may have to be answered on purely pragmatic grounds. If the procedure recognizes too many short and shallow fluctuations as cycles, the admission criteria must be tightened. This can be done in a variety of ways, for instance, by choosing less flexible (longer-term or differently weighted) trend-cycle approximations, by lengthening the span within which a peak is chosen or by extending duration minima. In order to choose between these tactical alternatives, it is necessary to analyze not only the nature of the contingencies that should be avoided but also the effects of alternative stratagems on the over-all efficiency of the chosen procedure. The question is whether any proposed alternative improves the process at large or only the results for a particular activity. This characterizes the approach and the choice between alternative criteria of selection as essentially heuristic.

In order to describe the selection process as well as the direction of desirable improvements, the experimental procedure currently in use will be described in some detail and illustrated by the monthly bituminous coal production series from 1914 to 1938. The computer output of the analysis is presented in the Appendix to this chapter.

[6] The complication would arise from the difficulty of setting adequate standards. Minimum amplitudes should be different for volatile and for stable activities, and thus presumably should be expressed in terms of average cyclical volatility for a given activity during reference cycle phases (since the determination of specific cycles now would hinge upon the volatility measure). To apply uniform standards, these averages should cover the same time periods. Moreover, the mild response of a given activity to a mild contraction in general business conditions may well be cyclically significant, irrespective of contraction amplitudes during other cycles. This raises the problem of comparison of cyclical amplitudes for any given activity with swings in business conditions at large or in representative activities—certainly no simple matter. Although it may be theoretically and technically feasible to incorporate explicit amplitude considerations in programmed turning point selection, such incorporation would complicate the procedures considerably and perhaps render them impractical.

The procedure will then be applied to a group of time series—to wit, the leading, coinciding and lagging business cycle indicators as reported by the Bureau of the Census of the U.S. Department of Commerce in its monthly *Business Cycle Developments,* now renamed *Business Conditions Digest* (B.C.D.). On the basis of the selected turns, the efficacy of the process will be evaluated and the needs for further work pointed out.

PROCESS OF SELECTION

Bituminous coal production, 1914–38, is a series with a good number of turning point problems, such as the presence of strong random and other irregular movements (strikes), of double turns, and of minor cycles. In addition, it is a series that has been analyzed and discussed in detail by Burns and Mitchell in *Measuring Business Cycles.* The seasonaly adjusted series, several smoothed versions, and tentative as well as final turning points are presented in Chart 3. The lowest curve on the chart shows the time series of seasonally adjusted data. The three other curves represent the results of several smoothing processes—a twelve-month moving average, a fifteen-month Spencer curve,[7] and a four-month moving average, respectively. These three curves are used in the gradual approximation of turns in the unsmoothed series. The essential steps of the procedure are outlined in Table 1.

EXTREME OBSERVATIONS

Since the representations of trend-cycle movements should be free of the influence of extreme observations, the identification of such observations and the derivation of suitable replacement values are the first steps in the program.

Extreme values are defined as values whose ratios to a fifteen-month

[7] This is a complex graduation formula, a weighted moving average with the highest weights in the center and negative weights at the ends, which ensures that the curve follows the data closely. Spencer curves can be considerably more flexible than an unweighted twelve-month moving average. This implies that the Spencer curve follows the original curve into peaks and troughs without drastic effects on the location of turning points—a valuable feature for a procedure of turning point selection. On the other hand, the flexibility of the Spencer curve causes it to follow minor fluctuations of less than cyclical importance and sometimes negligible amplitude. The latter feature may complicate the selection process, particularly if the procedure does not contain specifications for minimum amplitudes. Both curves are used in the present procedure.

CHART 3

BITUMINOUS COAL PRODUCTION AND MOVING AVERAGES, 1914–38

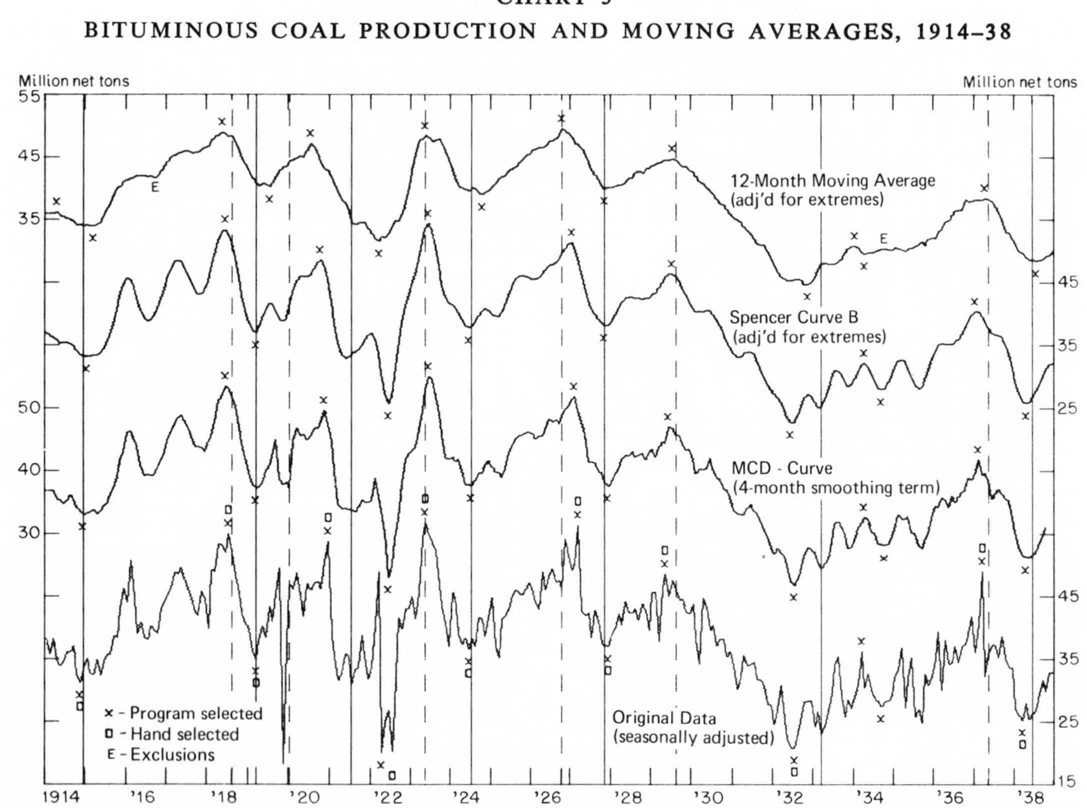

Note: Broken vertical lines denote business cycle peaks; solid vertical lines denote business cycle troughs.

TABLE 1

PROCEDURE FOR PROGRAMMED DETERMINATION OF TURNING POINTS

I. Determination of extremes and substitution of values.
II. Determination of cycles in 12-month moving average (extremes replaced).
  A. Identification of points higher (or lower) than 5 months on either side.
  B. Enforcement of alternation of turns by selecting highest of multiple peaks (or lowest of multiple troughs).
III. Determination of corresponding turns in Spencer curve (extremes replaced).
  A. Identification of highest (or lowest) value within ±5 months of selected turn in 12-month moving average.
  B. Enforcement of minimum cycle duration of 15 months by eliminating lower peaks and higher troughs of shorter cycles.
IV. Determination of corresponding turns in short-term moving average of 3 to 6 months, depending on MCD (months of cyclical dominance).
  A. Identification of highest (or lowest) value within ±5 months of selected turn in Spencer curve.
V. Determination of turning points in unsmoothed series.
  A. Identification of highest (or lowest) value within ±4 months, or MCD term, whichever is larger, of selected turn in short-term moving average.
  B. Elimination of turns within 6 months of beginning and end of series.
  C. Elimination of peaks (or troughs) at both ends of series which are lower (or higher) than values closer to end.
  D. Elimination of cycles whose duration is less than 15 months.
  E. Elimination of phases whose duration is less than 5 months.
VI. Statement of final turning points.

preliminary unadjusted Spencer curve (Spencer curve A) are outside a specified range. The present exclusion criterion is 3.5 standard deviations of the ratios, and is shown as "control limit = 3.500" on the title page of the output. The preliminary Spencer curve A is found in Output Table 2-2. The size of one standard deviation (7.853) is given at the bottom of this table, and the identification of extreme values is made in the subsequent lines. In the present case three values, all of them strike-related, are considered extreme: November 1919, March 1922, and April 1922—that is, all three of them deviate from Spencer curve A by more than 3.5 standard deviations. At the dates mentioned, the values of the unadjusted Spencer curve A are substituted for the extreme values in the original series in order to derive revised trend-cycle representations, i.e., Spencer curve B (not included in the output tables but presented in Chart 3) and a twelve-month moving average.[8] Note that July 1922, which has practically the same value as April 1922, is not excluded as extreme. The reason is that the unadjusted Spencer curve is much lower in July than in April (the value of which is strongly affected by the high extreme value of March), and thus the ratio of the July value to the Spencer curve value is less than 3.5 standard deviations from the mean of the ratios. In principle, an iterative procedure could lead to more consistent exclusions.

TURNS IN THE TWELVE-MONTH MOVING AVERAGE

The first curve from which turning points are determined, after adjustment for extreme values, is a twelve-month moving average (see Output Table 2-3 and the first curve in Chart 3). The reason for starting with the twelve-month moving average rather than with the Spencer curve is that the Spencer curve proved to be too flexible for our purpose (i.e., it contains too many minor fluctuations). The two curves can be compared in Chart 3. Most of the short fluctuations of the Spencer curve (1916, 1917, 1921, 1925–26, 1930–31, 1933, and 1935) are not reflected or only mildly reflected in the twelve-month moving average. Thus the latter curve seems to be a convenient means for eliminating fluctuations of subcyclical duration or of very shallow amplitude.

The selection of turning points is done in two steps: First, tentative turns are established, then these turns are tested for compliance with a set of constraint rules. Any month whose value is higher than those of the five preceding months and the five following months is regarded as the date of a tentative peak; analogously, the month whose value is lower than the five values on each side is regarded as the date of a tentative trough. In the case of bituminous coal production, the program picked a considerable number of such local maxima and minima, to wit, eight peaks and ten trough (see relevant output table). The turns selected on the twelve-month moving average are subjected to only one test—a check on the proper alternation of peaks and troughs. The elimination of multiple turns is simple. Of two or more contiguous peaks, the highest one (and if they have the same value, the latest)

[8] Experiments with alternative substitution rules, such as the replacement of extremes by the average of the nearest nonextreme values, led to the same or similar final results in practically all instances. Therefore, computations based on alternative substitutions were dropped from the procedure.

survives; and the analogous rule holds for troughs. In the present example, the excess troughs of September 1916 and April 1935 are removed (see Output Table 2-5). The remaining turns are marked by an X, the eliminated ones by an E in Chart 3.

TURNS IN THE SPENCER CURVE

The next step in the process is the determination of tentative and final cyclical turns in the Spencer curve. The Spencer curve is selected as the first intermediate curve because its turns tend to be closer to those of the original data,[9] a desirable step toward the final goal.

In principle, the program searches—in the neighborhood (delineated as ± five months) of the turns established for the twelve-month moving average—for like turns on the Spencer curve. That is, in the neighborhood of peaks, it searches for the highest of the eleven points on the Spencer curve; in the neighborhood of troughs, for the lowest. The Spencer curve turns thus located are subjected to two tests: (1) like turns must be at least fifteen months apart; and (2) the alternation of peaks and troughs must be maintained.

The stipulation that turns must not be closer than six months from the end of the series is, of course, introduced to avoid spurious highs or lows that have no cyclical significance. In the present illustration, the search did not turn up a Spencer curve peak that corresponded to the local maximum of April 1914 on the twelve-month moving average. This is expressed in the message "First turn is too near the beginning." Note that the search located a Spencer curve trough corresponding to the twelve-month moving average low of July 1938. The Spencer curve turns located by the described procedure are then listed.

The next test is designed to enforce a minimum-duration rule for recognized cycles. The adopted rule is that peaks as well as troughs must be at least fifteen months apart from like turns. After identifying like turns that are too close, the program excludes the lower of two peaks and the higher of two troughs. Exclusion of any turn requires elimination of an opposite turn to maintain the proper alternation of peaks and troughs. Sometimes this presents no problem. However, if there are several corresponding turns less than fifteen months apart,

[9] The equal-weight scheme of the twelve-month moving average can distort the location of turning points considerably. Compare, for instance, the turns in the twelve-month moving average, in the Spencer curve, and in the original data around the 1927 peak or the 1932 trough of the bituminous coal series.

a procedure must be used that will eventually lead to the elimination of several peaks and troughs. In the present example, no cycles on Spencer curve B are eliminated because of insufficient duration. This is almost entirely due to the use of the twelve-month moving average as a preliminary screening device. If the procedure had started with the Spencer curve, several of the short fluctuations mentioned before would have been provisionally recognized and later excluded under the fifteen-month duration rule.

The last test is designed to avoid "crossovers." If a contraction in the twelve-month moving average is less than ten months long, the searches for peaks and for troughs on the Spencer curve overlap. Hence the searches could conceivably lead to a Spencer curve contraction in which the low precedes rather than follows the peak. Since the conditions leading to such a crossover throw doubt on the existence of a genuine contraction, both turns involved are omitted.

The remaining turning points in the Spencer curve are listed in the output tables and were marked by us in Chart 3. It will be useful to review the efficacy of the procedure up to this point. On the whole, the delineated cycles seem reasonable; in particular, the omission of most of the briefer fluctuations should be regarded as successful. The only problem is the recognition of the brief 1934 contraction as cyclically significant, whereas the 1935 contraction is not recognized. The mechanics of the selection process are clear enough: The twelve-month moving average did not have a peak in the winter of 1934–35, and this eliminated April 1935 as a (multiple) trough. Thus the year 1935 did not fall into the search range of the program. The consequences of this restriction will be reflected in the final turning point determination, as will be seen later on.

IMMEDIATE NEIGHBORHOODS OF FINAL TURNS

It could be argued that the Spencer curve cycles should form the basis of cyclical analysis, since conceptually they are closest to the trend-cycle component of the observed values. However, as in all long-term moving averages, Spencer curves tend to shift turns, affect slopes, and convert irregular fluctuations into smooth wavelike patterns. Thus, analysis cannot be based on smoothed series alone, but must consider the behavior of unsmoothed observations.[10] Moreover, the exclusive use of smoothed series would not only make cyclical analysis depend-

[10] For a discussion of this problem, see Burns and Mitchell, *op. cit.*, pp. 310 ff.

ent upon the particular smoothing term and weighting scheme but would also be a radical departure from cycle measures previously used by the National Bureau and other investigators, and would impair comparability of research results. Cycles, as analyzed by the National Bureau, are based on unsmoothed values. Thus the search has to continue for values close to the Spencer curve turns that are peaks or troughs in the original seasonally adjusted data. This search could be carried out in the neighborhood of Spencer curve turns without use of any further intermediate curve, but there are possible drawbacks to such a procedure. The Spencer curve is a long-term moving average, quite capable of imparting a bell-type smoothness to data that form double or triple peaks or troughs (compare the curve contours in 1916, 1917, 1921, and 1935). Hence, the turns in the original data might be quite far from those of the Spencer curve, and consequently the procedure would require a correspondingly broad search range in order not to miss the turns. However, such a wide range would catch irregular maxima or minima that are not cyclically significant peaks and troughs. For this reason it was thought better to redelineate the neighborhood of the final turns by searching in the neighborhood of the Spencer curve turns for corresponding turns in a short-term moving average.

A curve that represents the original seasonaly adjusted data smoothed by a short-term average is the MCD curve. MCD stands for months of cyclical dominance. The MCD of any series is the number of months required for the systematic trend-cycle forces to assert themselves against the irregular time series component. If a series has strong cycles and little irregularity, it will not take long (perhaps not longer than one or two months) until the average change in the trend-cycle component exceeds the average change in the irregular component. If a series has shallow cycles but is very choppy, it may take many months before the cyclical movement asserts itself. In the first case no smoothing, or smoothing by only a very short-term average, is required to bring out the cyclically relevant movements; in the second case a correspondingly longer term is needed.[11] The MCD curve is the curve representing the data smoothed by the MCD term appro-

[11] Technically, the number of months required for dominance of the cyclical over the irregular component is that span over which the average change in the irregular component becomes smaller than the average change in the trend-cycle component. For further explanation, see Julius Shiskin, "How Accurate?" *American Statistician,* October 1960, pp. 15 ff.

priate for the relationship of trend-cycle and irregular movements in the analyzed activity. In order to reduce the effect of irregular changes and the influence of remote values, the span of the smoothing term used is confined to the narrow range of three to six months. For measured MCD's of one and two months, a three-month term is applied; for MCD's of seven or more months, a six-month term is used.

The MCD for bituminous coal production is four months. The MCD curve is plotted as the third curve of Chart 3 and is based on Output Table 2-4. The method of deriving turning points in the four-month moving average is practically the same as that described in the preceding section. Turns are determined by selecting the highest peak on the MCD curve within a span of five months from the dates of the peaks on the Spencer curve; MCD troughs are analogously selected. Before this determination is made final, turns at the very beginning and end of the MCD series are omitted, the minimum duration rule is enforced, and the turns are tested for crossovers. In the present case, no further exclusions result from the application of these tests. The remaining turns are reported in the output table as turning points in four-month moving average; they are marked by crosses on the third curve of Chart 3.

SELECTION OF FINAL TURNING POINTS

The last step of the procedure is to find the peak and trough values in the unsmoothed data that correspond to the MCD turns previously established.[12] This simple search is analogous to the previous transitions (from turns in the twelve-month moving average to Spencer curve turns and from Spencer curve turns to MCD curve turns). The program establishes the highest values in the unsmoothed data within a span of ± MCD or ± four months (whichever is longer) from the peak in the MCD curve; correspondingly, the lowest value of the unsmoothed data in the neighborhood of MCD troughs is established. No turns closer than six months from the ends are accepted. Also, the first and last peak (or trough) must be at least as high (or as low) as any value between it and the end of the series. The resulting turns are reported in Output Table 2-8.

[12] Again it could be proposed that the dates and values of the MCD turns should be regarded as those relevant for cyclical analysis, since the unsmoothed series is modified by irregular elements that are not intended to affect cyclical measures. Whatever the merits of this view, the present analysis ignores it in order to adhere as closely as feasible to standard practices.

The final tests deal with the duration of cycles and cycle phases. Full cycles (peak-to-peak and trough-to-trough) are checked for a minimum duration of fifteen months. The fact that this criterion was applied earlier to the trend-cycle curve does not necessarily mean that the related cycle in the unsmoothed data satisfies the condition, as the actual initial and terminal turns can be closer than the related turns on the Spencer curve. In the present example, however, the application of the test does not lead to further exclusion of cycles. The last constraint for which the turning points are tested is a five-month minimum rule for phase durations. There is no equivalent rule in the standard analysis of the National Bureau. However, early experiences showed that some sharp, short episodic movements (such as strikes) can be redistributed by the various moving averages into fluctuations of cyclical contours and durations. The minimum-phase-duration rule, which at present is set at five months, is a possible remedy.[13] This rule could be readily modified if experimentation or extended experience should prove it to be inadequate.

The final turning points are listed on Output Table 2-8 and marked by crosses on the lowest curve of Chart 3. Comparison with those previously selected,[14] and marked by squares on the same curve shows one minor and one major discrepancy. The minor difference consists in the selection of April 1922 instead of July 1922 as a trough. The situation here is that of a characteristic double trough. The earlier trough, in April, is slightly lower and thus was selected under the specified procedures.[15] The difference is unimportant and could be passed over without comment were it not that it illustrates some characteristics of programmed turning point determination. The National Bureau's selection of the July turning point is explained as follows:

> The trough in 1922 exemplifies a "double bottom." There is a deep trough in April 1922, when a strike—probably the greatest in the history of this afflicted industry—broke out. A slight revival occurred during the next two months, and a relapse in July, when the railroad shopmen's strike produced

[13] Different minima for expansions and contractions were considered, in view of the longer durations of expansions in historical business cycles, but were ruled out in order to make the program equally effective for series with positive and inverted conformity and for series with rising and falling long-term trends.

[14] Burns and Mitchell, *op. cit.*, pp. 60 ff.

[15] April 1922 was identified above (p. 21) as an extreme value. This does not prevent it from being chosen as a turning point. Prevailing practice is to accept extremes as turning points if they occur in a turning point neighborhood.

an acute car shortage in the non-union field. The seasonally adjusted figure is fractionally higher in July than in April (20.2 against 20.1 million tons). But the difference is negligible, and in line with our rules, the trough is dated in the later month.[16]

Apart from the consideration of historically unique events, such as the two strikes, it would be difficult to program a provision to neglect "negligible" differences. When is a difference negligible? Should the same standard be applied to all series, whatever their volatility? And if there are several alternative troughs, each negligibly different from the other but arranged in ascending order, should the last one nevertheless be selected? While it is not technically impossible to incorporate these and similar considerations in a programmed selection, it would lead to a proliferation of tests that might make the process unwieldy and perhaps impractical—at least at the present state of the art.

The major discrepancy between the traditional and the programmed turning point determination is the recognition by the programmed procedure of a cyclical contraction in 1934. The problem is not only the debatable recognition of any contraction at all but the specification of the contraction as lasting six months, from March to September 1934, instead of lasting for one more year, to September 1935. The latter would be the more plausible version, in view of the behavior of the actual data. The technical reason for the programmed determination was discussed before in connection with the turning point selection on the Spencer curve. It goes back to the use of the twelve-month moving average as the first step in the process of cycle identification. It would, of course, be quite simple to *increase* the flexibility of the first average, decrease the span, and thus permit consideration of the events of 1935. Alternatively one could *decrease* the flexibility of the first curve, and thus eliminate the cycle in question altogether. Whatever the solution to the problem, its formulation and acceptance cannot be based on the analysis of programmed turning point determination for a single series or even for a few series. It is always possible, and relatively easy, to modify the program to cover a small number of contingencies. The goal is to develop rules that operate in a satisfactory manner for most economic time series. Thus, experimentation with a fairly large sample of series must be resorted to in order to establish the efficacy and the shortcomings of a given set of rules.

[16] Burns and Mitchell, *op. cit.*, p. 63.

Analysis of programmed turning points and modification of procedures must be based on experiences with such a sample. This is the concern of the following section.

EMPIRICAL EVALUATION OF PROCEDURE

DESCRIPTION OF SAMPLE

The sample chosen for the experiment is the collection of leading, lagging, and coinciding business cycle indicators for the period starting 1948, as published monthly in *Business Cycle Developments*.[17] There are several reasons for this choice. First, the sample covers a large number of important activities representing various aspects of economic life in the United States. Second, the selected series exhibit a great variety of behavior: there are series with relatively large random components and little trend, such as temporary layoffs; there are smooth series, with little random and strong upward trend, such as personal income; there are positively conforming series, and there are inverted series; there are monthly and quarterly series; there are series with and without negative entries. Third, the series were all available, in up-to-date and seasonally adjusted form, in *Business Cycle Developments,* thus reducing or eliminating the chores of data collection and seasonal adjustment. Fourth, and for present purposes most important, cyclical turning points have already been established by the National Bureau for all these series, based on the rules given by Burns and Mitchell in *Measuring Business Cycles*. This makes comparisons of previously selected and program-selected turns possible for all series. *Business Cycle Developments* reports thirty leading, fifteen coinciding, and seven lagging indicators, or fifty-two altogether. One series, new approved capital appropriations (series 51), had to be omitted because of several discontinuities. Thus, fifty-one series were left for experimentation—a sufficiently large and variegated sample for present purposes.

Despite its broad coverage, the sample described has one serious bias that may affect its value for testing the broad applicability of programmed turning point determination for economic time series in general. The bias arises from the fact that most of the series shown in *Business Cycle Developments* were chosen because their marked

[17] The sample used was published in *Business Cycle Developments* before the changes instituted in the April 1967 issue.

cyclical characteristics made the series valuable for diagnosis of current business conditions. That is, most of these series have recognizable if not pronounced cycles, show good conformity to several business cycles, and are not excessively affected by random elements. The greatest difficulties in turning point determination arise when cyclical components are weak and irregular factors are strong—and this situation is rare in the selected sample. However, at the present stage of development the programmed approach cannot be expected to solve problems that proved intractable before. Thus, although the sample excludes series whose cyclical behavior is particularly uncertain, it will serve well to test the broad applicability of the approach and to suggest the direction in which further progress should be sought.

CRITERIA FOR EVALUATION

Before reporting the outcome of the experiment, it might be well to consider the criteria by which the results should be evaluated. An obvious standard for evaluation is the turning point selection that had previously been made for these series. Although for many reasons—particularly comparability with previous work—this appears to be a desirable standard, it is not without weaknesses.

For one thing, the general rules of turning point determination are subject to interpretation; and, if a basic rule permits more than one choice, the choice actually made in the past should not serve as criterion for the evaluation of the programmed selection. Second, in the selection and analysis of cyclical indicators, doubtful cases might have been resolved by accepting fluctuations that conformed to fluctuations in business activity at large.[18] Thus, neither the programmed selection of additional intraphase turns nor the omission of conforming but otherwise doubtful phases should necessarily be regarded as a defect of the program. Third, it is possible that the technology of programmed selection requires somewhat different rules than those developed as guides for the exercise of judgment by individuals. Thus, there exists the possibility that the ground rules may have to be changed to facilitate programmed selection. Such changes should, of course, be made only if clearly justified by the results of broad experimentation.

The preceding remarks should not be interpreted as a defense of

[18] Burns and Mitchell, *op. cit.*, p. 58.

the programmed selection of turns whenever these depart from those previously determined. After all, the program is experimental and excludes from consideration certain criteria, such as comparative amplitude and runs of changes in the same direction, which proved to be valuable guides in the past. It is precisely the purpose of this report to establish whether programmed selection can safely ignore some of these guides or whether the program must be amended to reflect these and other considerations. The most important criterion for the general usefulness of the program is whether basic research findings are affected by its use.

PROGRAM-DETERMINED AND STAFF-DETERMINED TURNS

In Chart 4, the turns picked by the program are marked by crosses placed close to the line and those chosen previously by the Bureau are marked by squares. Even a casual examination of the chart reveals that in most instances the program picked the same points as those established earlier, but that the program often recognized a number of short and mild fluctuations as cyclical that were not so regarded in previous analyses. The opposite situation, that of fluctuations recognized previously but not by the program, occurs less frequently. In comparing the results, a distinction must be made between the identification of specific cycles, and the precise dating of their turns. The first implies the recognition of certain neighborhoods as turning point neighborhoods; the second involves specification of the month in which the turn occurred. Let us begin with the question of recognition.

In the following analysis, interest is concentrated on the differences between the results of programmed and nonprogrammed determination of turning points. Table 2 shows that the program found 432 specific phases, as against 384 found by the nonprogrammed approach. Identical phases were found by both approaches in 346 cases. That means the program found 90 per cent of the phases previously established by the National Bureau staff. Since the program found 48 more phases than the Bureau, the phases found by both approaches constitute only 80 per cent of all phases determined by the programmed approach.

The net difference of 48 does not provide a satisfactory criterion for evaluating the similarity of the results. The net difference hides the fact that there are 124 phases that were chosen by one approach

CHART 4

BUSINESS CYCLE INDICATORS, TURNING POINTS SELECTED BY PROGRAMMED AND NONPROGRAMMED APPROACH, 1947–67

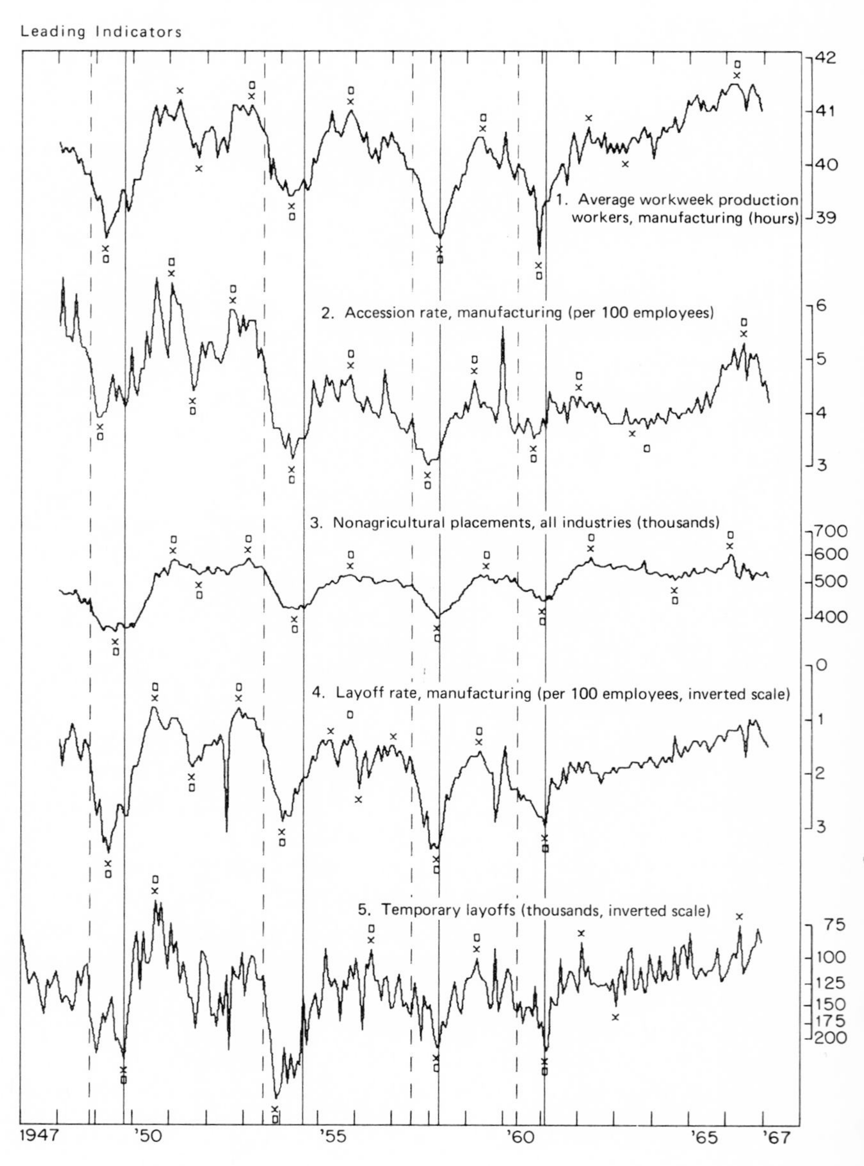

CHART 4
(Continued)

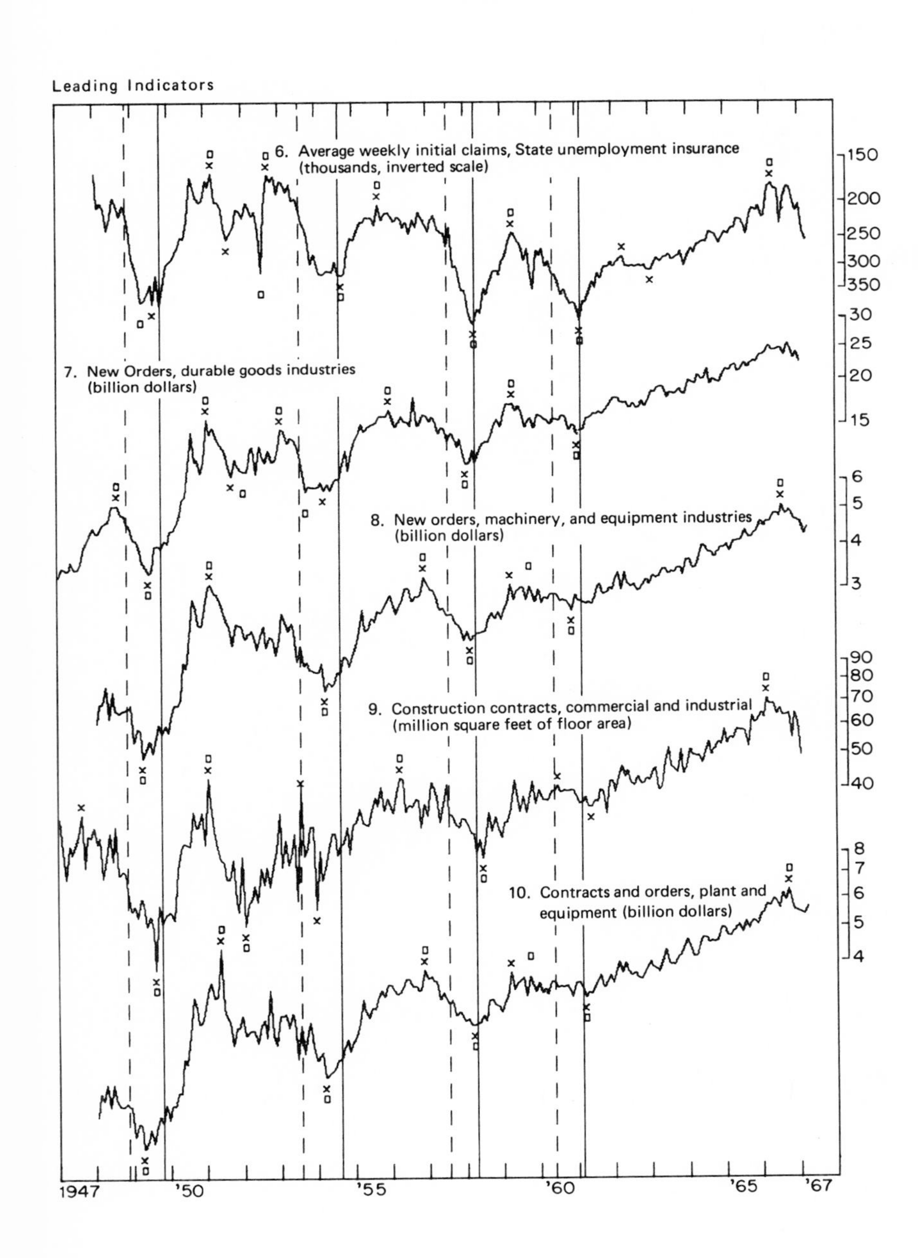

CHART 4
(Continued)

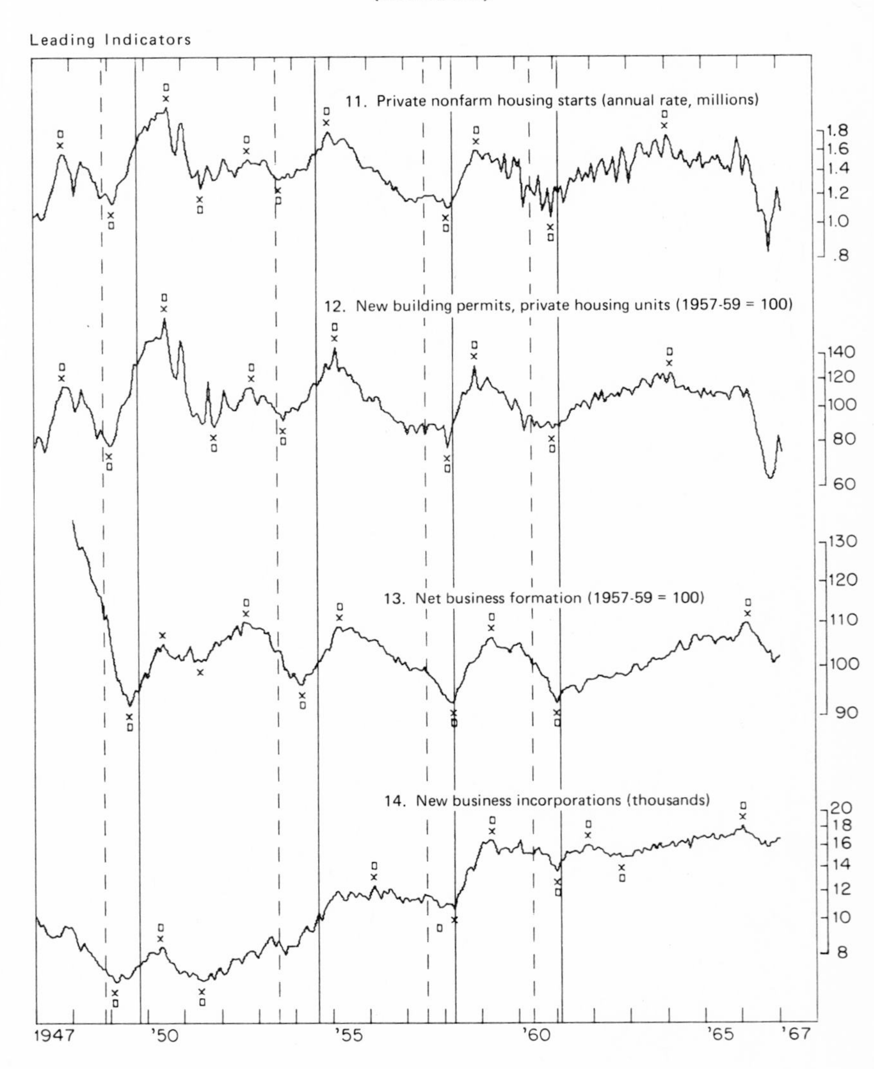

CHART 4
(Continued)

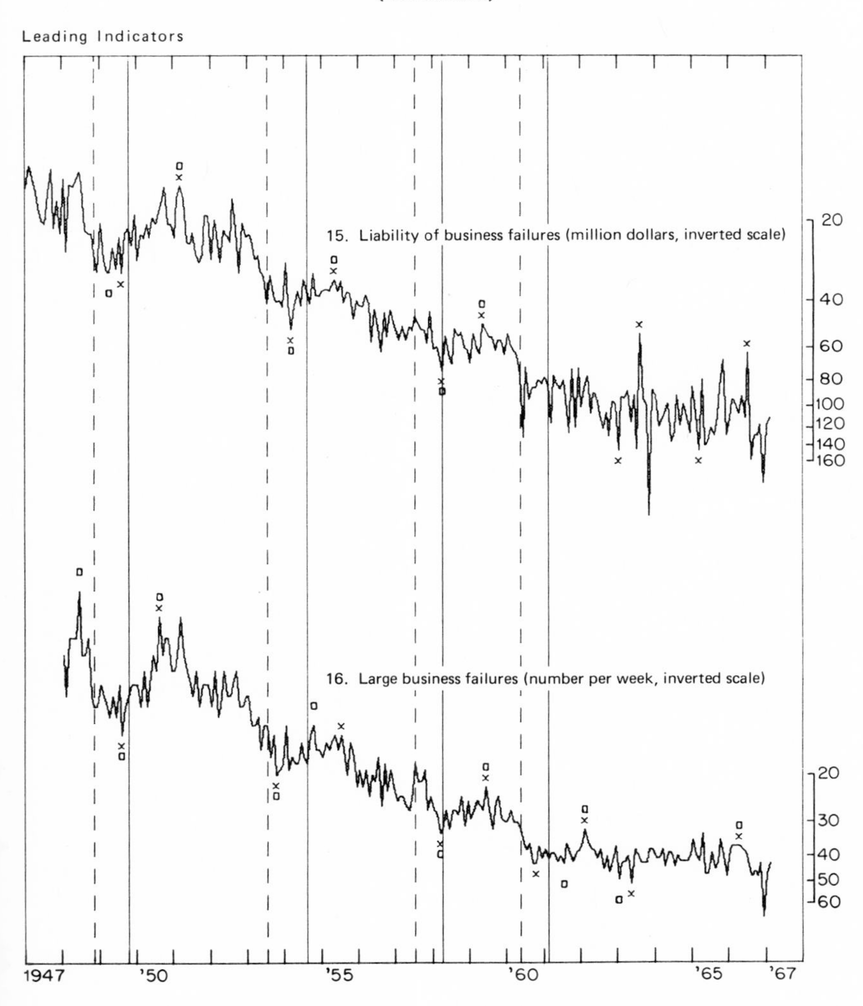

CHART 4

(Continued)

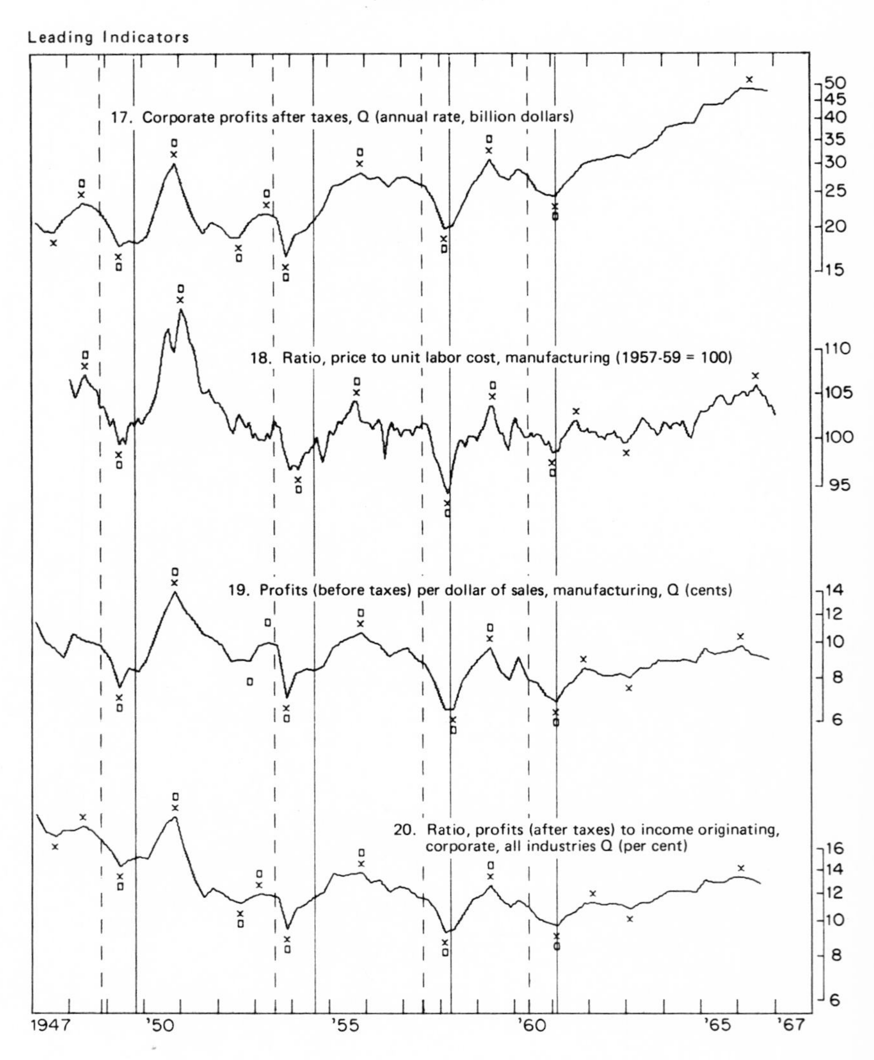

CHART 4
(Continued)

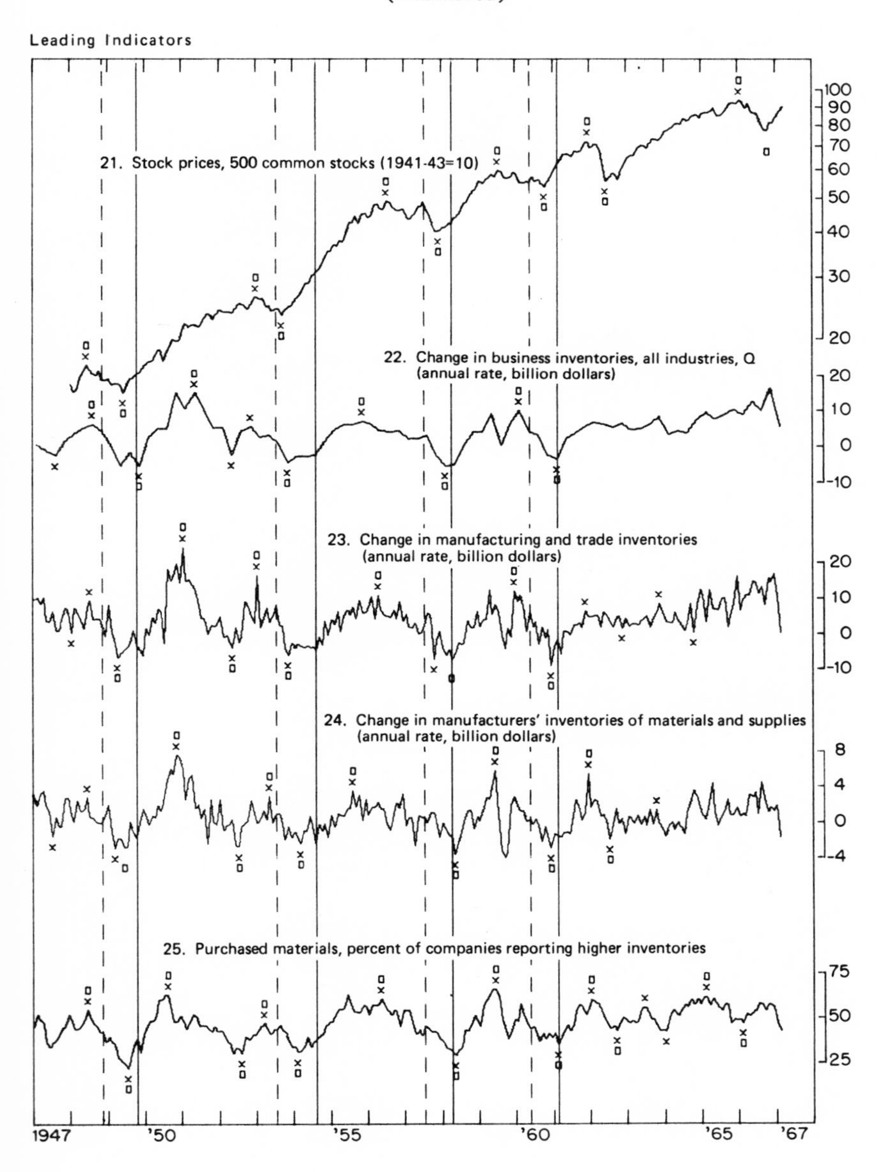

CHART 4
(Continued)

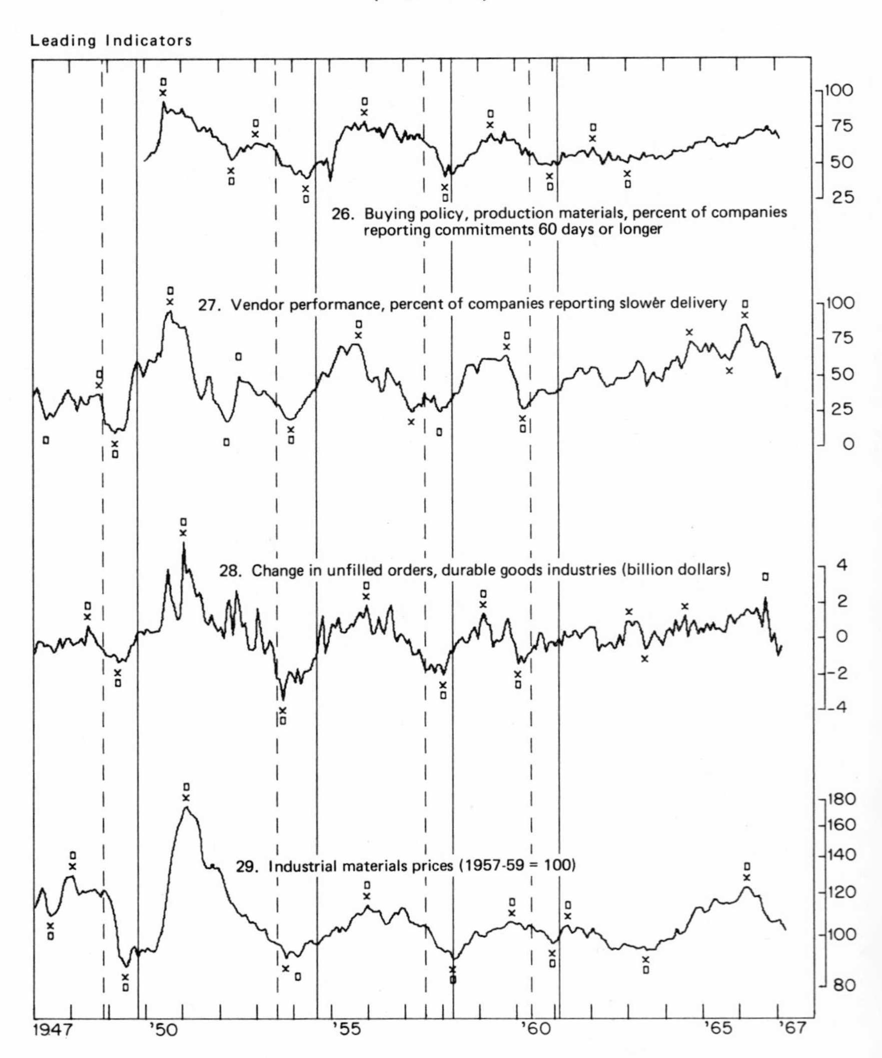

CHART 4
(Continued)

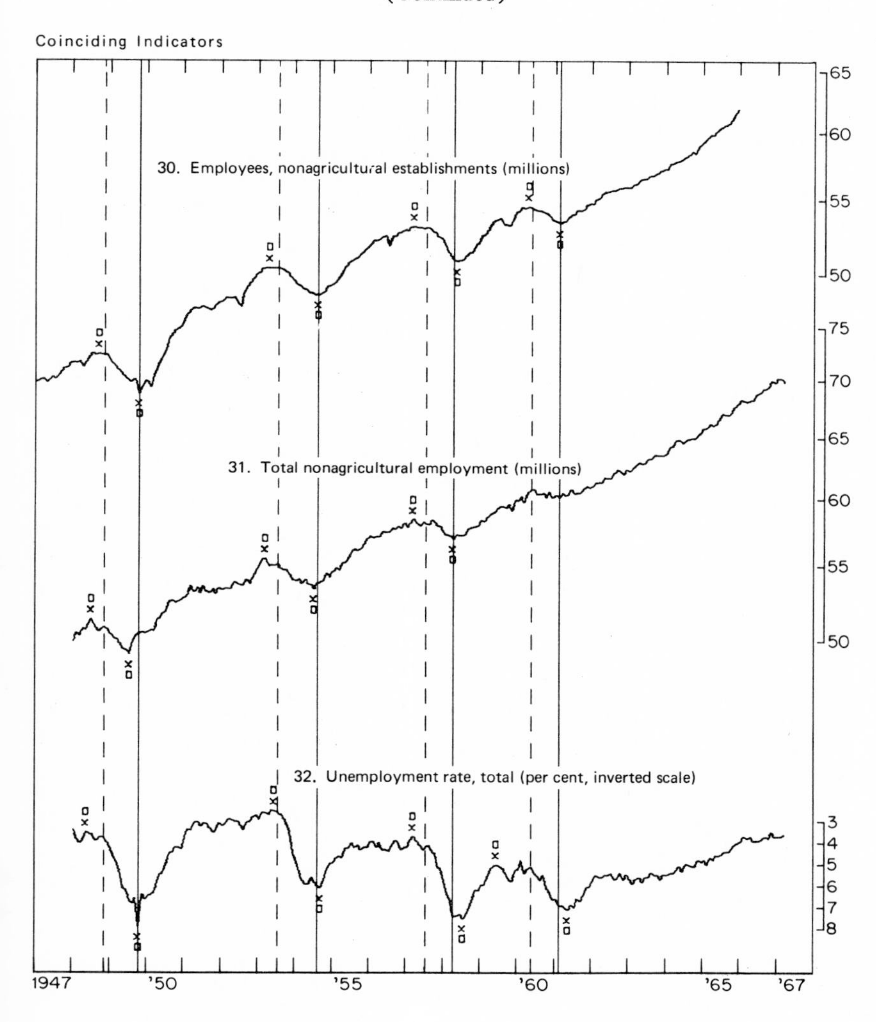

CHART 4

(Continued)

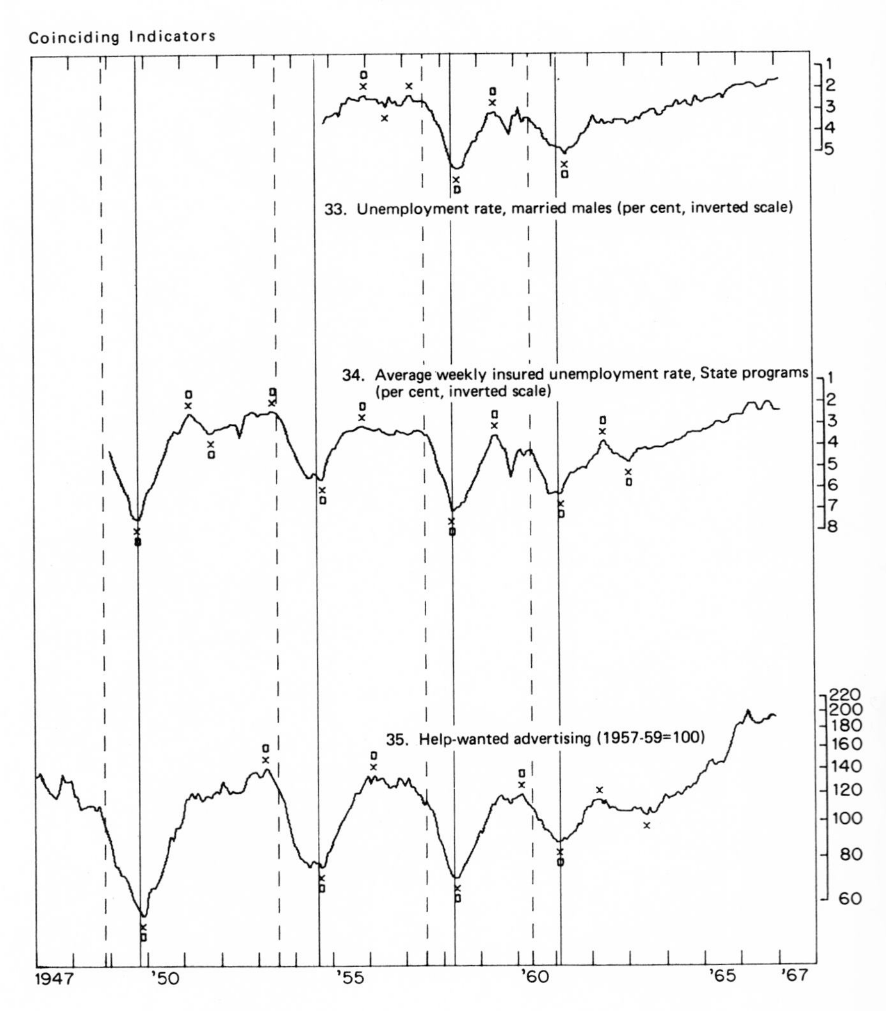

CHART 4
(Continued)

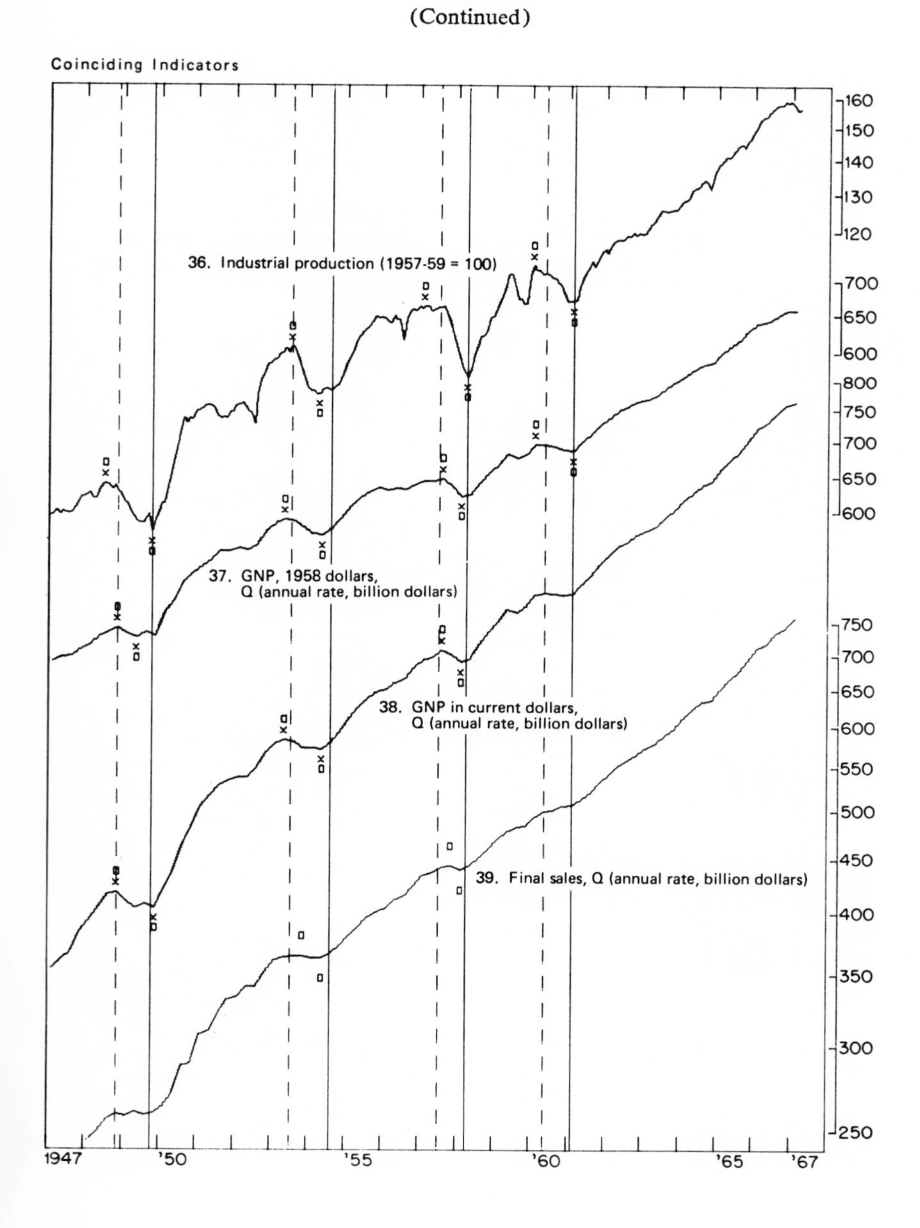

CHART 4
(Continued)

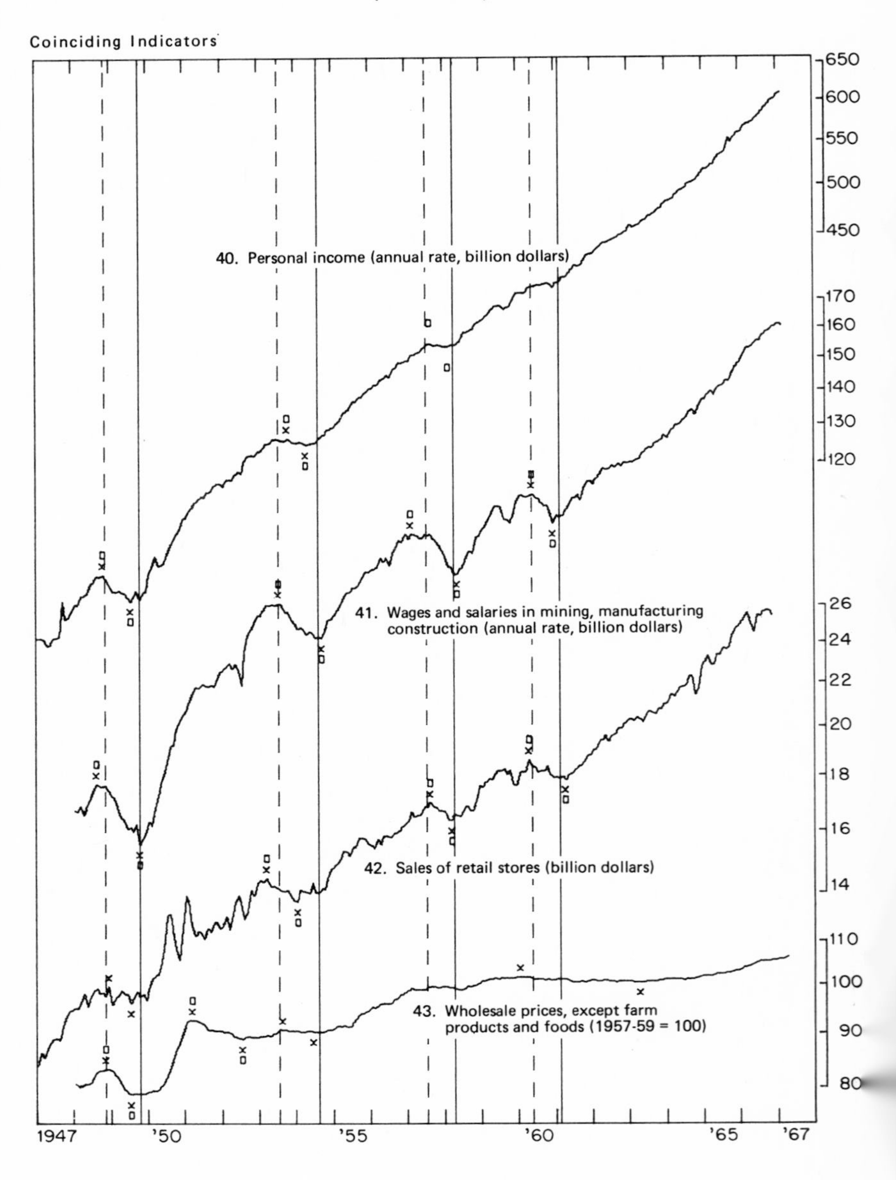

CHART 4

(Continued)

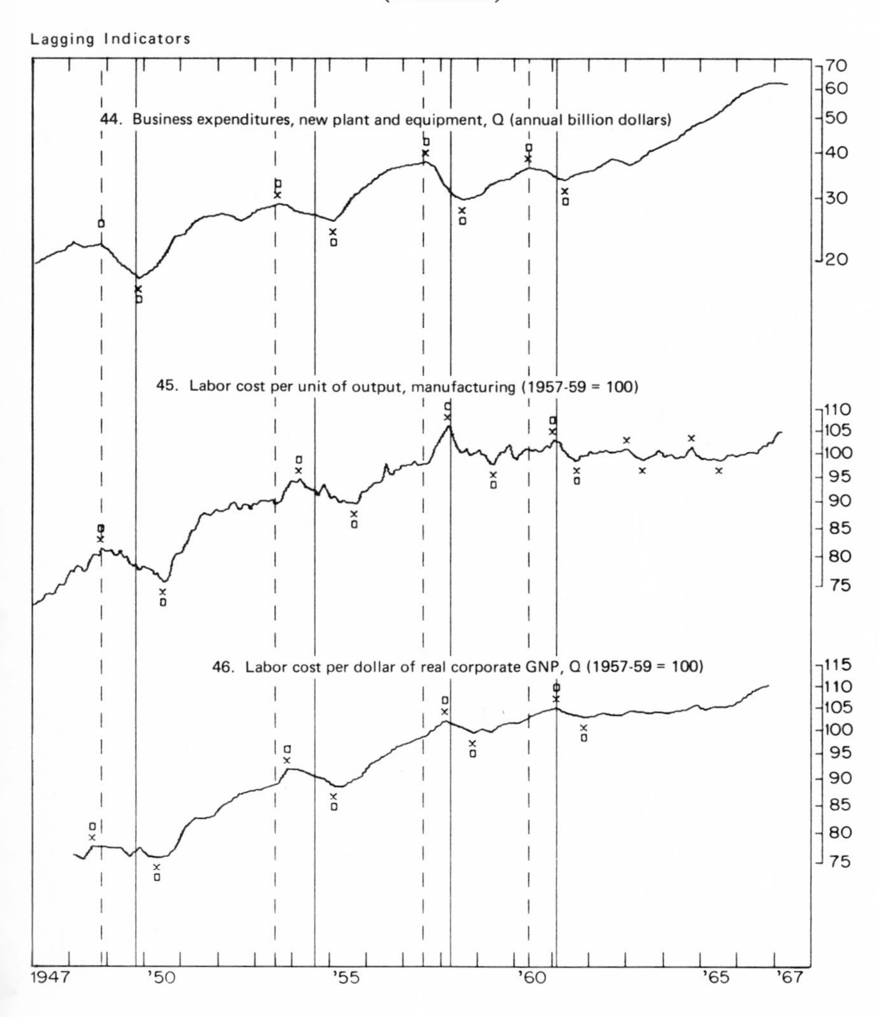

CHART 4

(Concluded)

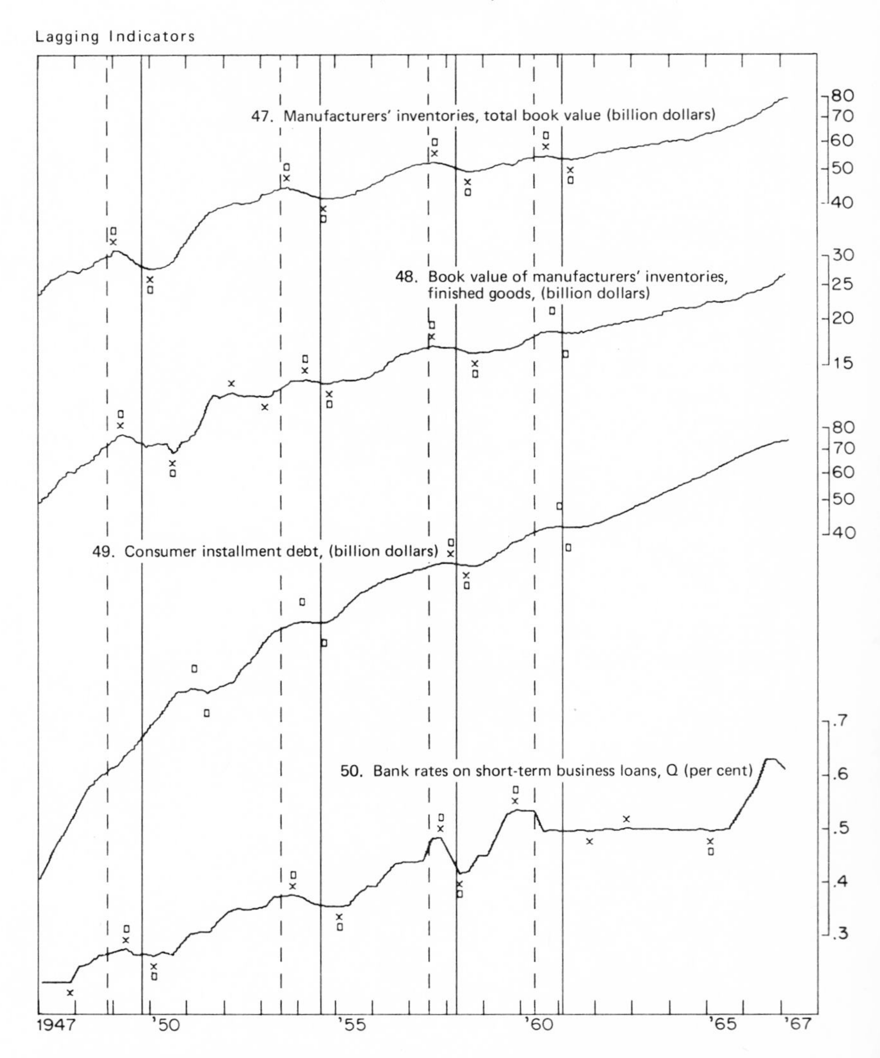

Note: Crosses denote program determined turning points of the specific series; squares denote turning points of the specific series previously determined by the National Bureau staff. Broken vertical lines denote business cycle peaks; solid vertical lines denote business cycle troughs.

TABLE 2
SPECIFIC-CYCLE PHASES
IN BUSINESS CYCLE INDICATORS,
PROGRAMMED AND NONPROGRAMMED APPROACHES,
1947–67

	Expansions	*Contractions*	*All Phases*
1. All specific phases found by programmed approach	214	218	432
2. All specific phases found by nonprogrammed approach	186	198	384
3. Specific phases found by both programmed and nonprogrammed approach[a]	166	180	346
4. Specific phases found by programmed but not by nonprogrammed approach	48	38	86
5. Specific phases found by nonprogrammed but not by programmed approach	20	18	38
6. Phases found by both approaches as a percentage of those found by nonprogrammed approach	89	91	90
7. Phases found by both approaches as a percentage of those found by programmed approach	78	84	80

Source: Chart 4.
[a] Phases with corresponding initial and terminal turning points only.

but not by the other—86 by the program but not by the National Bureau staff, and 38 by the staff but not by the program. On the other hand, the differences shown in the table overstate the differences in the results of the two approaches. Phases are regarded as corresponding only if both the initial and the terminal turning points correspond. That is, if the program finds an intermediate contraction during a cyclical upswing of the nonprogrammed approach, this is counted as four differences—three extra phases found by the program and one extra phase found by the National Bureau staff. An example can be found to illustrate this point in the first expansion of the average workweek series, the first series shown in Chart 4. Here the contraction found by the program in 1951 gives rise to the report of four noncorresponding phases, i.e., of the three phases found by the program (1949–51, 1951–51, 1951–53) and the noncorresponding expansion found by the National Bureau staff for 1949–53.

TABLE 3

CYCLICAL COUNTERPHASES IN BUSINESS CYCLE INDICATORS, PROGRAMMED AND NONPROGRAMMED APPROACHES, 1947–67

Specific phases found by both programmed and nonprogrammed approach			346
Counterphases found by programmed but not by nonprogrammed approach			40
Corresponding to business cycle phases		8	
Not corresponding to business cycle phases		32	
Corresponding to business cycle retardations	19		
Counterphases found by nonprogrammed but not by programmed approach			14
Corresponding to business cycle phases		8	
Not corresponding to business cycle phases		6	
Corresponding to business cycle retardations	1		

Note: Counterphases are extra expansions (contractions) according to one approach, while the series experienced a contraction (expansion) according to the other approach. Extra turns, found by an approach at either end of a series, imply counterphases; these are included in the numbers given above.

In order to avoid the quadruple counting, Table 3 is introduced. This table reports only counterphases, that is, contractions (or expansions) that are found by one approach during expansion (or contraction) phases found by the other. In the fluctuations of the average workweek between 1949 and 1953, only one counterphase occurs—the extra contraction found by the program during 1951. The summary shows that the program found only forty counterphases. Moreover, eight of these correspond to concomitant business cycle phases and nineteen to the major retardations in business activity that occurred during 1950–51, 1962–63, and 1966–67.[19] Thus, most of the counterphases found by the program are economically plausible. The counterphases found by the nonprogrammed approach amounted to only fourteen, with eight phases conforming to corresponding business cycle phases and one to the retardation of 1950–51. This means that,

[19] These neighborhoods were singled out since many leading and a fair number of coinciding indicators showed declines, or at least retardations.

with only a few exceptions, the National Bureau staff agreed with the cyclical direction implied in the programmed determination of specific cycle phases.

When additional phases, such as the specific contraction from 1952 to 1953 in manufacturers' inventories of finished goods (series 48), are recognized by the program, the basic reason is simply that the twelve-month moving average went down long enough to establish a high and a low, each in the center of an eleven-month period. When a contraction recognized by inspection is not chosen by the program, as is the case in the specific decline from 1960 to 1961 in the same series, the reason is that the twelve-month moving average did not go down or did not go down long enough to qualify the decline as a cyclical contraction under the adopted criteria. It is true that other rules (alternation of peaks and troughs, minimum cycle length, minimum phase length, etc.) affect the final selection of turns, but the behavior of the twelve-month moving average controls basic eligibility. One case where the program omitted a phase selected by the National Bureau staff—because of insufficient duration, although the twelve-month moving average shows cycles—is the 1952 expansion in vendor performance (series 27).

The cited examples of differences in cycle recognition, and others that could be easily adduced, raise the question whether the programmed or the previously determined cycles are analytically preferable. This question is hard to answer without formal standards or at least some guiding considerations. In the case of manufacturers' inventories (series 48), the program-selected extra contraction of 1952–53 exceeds the extra contraction of 1960–61 selected by the Bureau staff in length and in amplitude and thus seems a better choice. The Bureau staff's recognition of the 1960–61 movement as a cyclical decline was presumably influenced by the fact that it corresponds to a business cycle contraction. On the other hand, the program's recognition of an extra contraction during 1956–57 in the layoff rate (series 4) seems inferior to the judgment of the staff member who regarded 1955–58 as one long specific expansion.[20]

[20] The program's recognition of the additional contraction was partly a consequence of unsatisfactory dating of turns—a point which will be discussed later on. Proper dating of the trough of the (inverted) layoff rate in November 1955 would have ruled out recognition of the ensuing increase as an expansion because the period of increase would have been below minimum phase duration.

Only five of the forty counterphases found by the program must be characterized as incompatible with a reasonable interpretation of the basic rules. The counterphases shown by series 1 in 1962–63, series 15 in 1963–63, series 33 in 1956–57, series 42 in 1948–49, and series 50 in 1961–62 are minor movements in comparison with the typical cyclical variations exhibited by these activities. All other extra-phases consist of mild fluctuations with fairly clear cyclical characteristics. The acceptability of such mild movements as cyclical phases depends on research goals and perhaps on the economic characteristics of experiences during the historical period analyzed. If one is interested in the timing and the degree of synchronization of fluctuations in economic activities during recent years, one may have to recognize mild fluctuations since they are the only ones present. In fact, analytical interest is shifting to the timing of changes in growth rates, so that analysis of fluctuations is not limited to actual declines in the level of activities. On the other hand, recognition of mild fluctuations may well be less desirable—or less important—for research concerned with economic fluctuations during the period before World War II. In principle, the program could be modified to accommodate these differences in objectives and historical context. However, such modification would diminish the procedural stability necessary for a uniform derivation of cyclical turning points and would thus be undesirable.

The above comparisons dealt with the recognition of cycle phases. Table 4 deals with the comparison of program-determined and staff-selected cyclical turns for leading, coinciding, and lagging indicators. Altogether, the difference between the results of the two approaches amounts to about 20 per cent of all phases, with a clear tendency of the program to pick more turns.

The program found 483 turns in the series as compared with 435 selected by the Bureau staff, for a net difference of 48 turns, or 11 per cent of the previously selected turns. These figures, however, overstate the agreement of the two selections since the program picked 72 turns where no corresponding turns were recognized by the National Bureau staff, and the staff picked 24 turns where no corresponding turns were found by the program. The resultant sum of 96 discrepancies in turning point recognition is double the size of the net differences. The program tended to recognize more cyclical turns both in leading and coinciding indicators, but particularly in the leading group with its more volatile activities. Altogether, the 411 corresponding

TABLE 4

CYCLICAL TURNING POINTS IN BUSINESS CYCLE INDICATORS, PROGRAMMED AND NONPROGRAMMED APPROACHES, 1947–67

	Indicators			
	29 Leading	*14 Coinciding*	*6 Lagging*	*All Indicators*
1. All turns found by programmed approach	324	103	56	483
2. All turns found by nonprogrammed approach	281	98	56	435
3. Corresponding turns found by both programmed and nonprogrammed approaches	272	92	47	411
a. Identical dates	255	92	47	394
b. Different dates	17	0	0	17
4. Noncorresponding turns				
a. Found by programmed but not by nonprogrammed approach	52	11	9	72
b. Found by nonprogrammed but not by programmed approach	9	6	9	24
5. Corresponding turns as a percentage of those found by nonprogrammed approach				94
6. Identical turns as a percentage of those found by nonprogrammed approach				91
7. Corresponding turns as a percentage of those found by programmed approach				85
8. Identical turns as a percentage of those found by programmed approach				82

turns constitute 94 per cent of all turns found by the nonprogrammed approach and 85 per cent of the more numerous turns found by the programmed approach.

While there are systematic differences between programmed and unprogrammed cycle recognition, differences between the dates of comparable turns selected by the two approaches are of minor importance. This can be seen in Chart 4 and lines 3a and 3b of Table 4. Among 411 cyclical turns that are recognized both by the program and by inspection, the program picked the same date in 394 turns (96 per cent of all corresponding turns); this represents 91 per cent of all staff-selected turns and 82 per cent of all turns found by the programmed approach.

In 12 of the 17 turns with different dates—which all occurred in the leading indicators—the National Bureau staff picked later turns than the program, in keeping with the Bureau's preference for resolving doubtful cases in favor of the later turn. Only four program-established dates are clearly inferior. They occur in the following circumstances:

Series	*Activity*	*Type of Turn*	*Date of Program-Selected Turn*	*Date of Staff-Selected Turn*	*Cause for Program Selection*
2	Accession rate	T	June 1963	Nov. 1963	Lower Spencer curve
4	Layoff rate	T[a]	May 1955	Nov. 1955	Lower Spencer curve
6	Initial claims, state unemployment insurance	P[b]	Aug. 1949	Apr. 1949	Higher peak
27	Vendor performance	T	Mar. 1957	Dec. 1957	Lower Spencer curve

[a] Shown as peak on inverted scale.
[b] Shown as trough on inverted scale.

In three of the four instances the basic cause for the discrepancy was that the Spencer curve was lower in the neighborhood of the program-selected trough; and the staff-selected trough (which was lower and/or later in the unsmoothed series) was beyond the stipulated search range. In series 2, for example, the program picked a turn in the middle of the flat-bottom trough of 1963 although there is a lower point at the end of the year—the turn picked by the National Bureau staff. The reason is that the Spencer curve has its trough in the beginning of the year, which puts the December low of the three-month moving average and the November low of the unsmoothed series beyond the respective search ranges. The search range, from Spencer to short-term average, would have to be greatly increased to catch the late 1963 low. However, such an increase would also result in a recognition of the irregular high in December 1959 as a peak, a clearly undesirable result. The limited range of the search leads to a questionable selection also in case of the peak (shown on the chart as a trough, due to inverted scale) of initial claims (series 6) in 1949. Here the program chooses the middle value of three prongs although the last value is higher and the early value, chosen by the staff, is better supported by adjacent values. The programmed approach did not search as far as the last prong and did not choose the first because it always picks the outlying value of the unsmoothed data without regard to adjacent values.

Application of the process to other series than those used in the present study revealed a potential weakness. In excluding turns associated with "short" phases (less than 5 months), the alternative peak selected by the program may be lower than the eliminated one, or the alternative trough may be higher. This result would be justified if the eliminated turn were randomly high or low but not if it reflected a cyclical reversal. Further experience may lead to program modification; in the meantime the user should be aware of the problem.

On the whole, our experience suggests that programmed turning point determination will prove useful for many research purposes. While the program-determined turns may be inferior to those established by experienced research workers, they may well be superior to those found by nonspecialists. The program is objective with regard to procedure; thus, the same turns will be found by every investigator who relies upon the program.

One research result that would not be significantly altered, whether the programmed or staff-selected turns are used, is the classification of economic indicators according to timing characteristics. Also, since program-determined and staff-selected dates are identical for all coinciding indicators, it must be presumed that the dating of reference cycles (which depends heavily on specific cycle turns in the coincident series) would not be substantially modified by the adoption of a programmed selection of turning points. However, the propensity of the program to pick up the relatively mild fluctuations, which characterize recent experience, affects measures of average cycle durations, overall amplitudes, and so forth.

APPLICATION TO REFERENCE TURNS

In the preceding section it was established that the classification of individual indicators into groups of leading, lagging, and coinciding measures would not be affected if turns were established by the programmed approach rather than by inspection. This, of course, does not imply that identical turns are selected by the two methods.

In this section we wish to establish whether and to what extent measures, such as cumulative historical diffusion indexes,[21] are affected by the substitution of program-determined turning points for staff-selected ones. A particularly important aspect of this question relates to summary measures of coinciding indicators, since their turns are strategic determinants of reference turning points, i.e., of the benchmark dates chosen to identify peaks and troughs in business activity at large. If it should turn out that cyclical turning points in cumulative diffusion indexes of coinciding indicators, based on program-determined specific turns, conform well with turns established by inspection, then the programmed approach may become a tool for reference turn determination. Finally, since the program tends to select more cycles than does the previous Bureau approach it would be interesting to establish whether these additional cycles—similar to cycles corresponding to those selected by inspection—are sufficiently synchronized to lead to recognizable swings in the cumulative diffusion indexes; and if so, whether these extra swings are related to known periods of business retardation.

The evidence is presented in Chart 5 and Table 5. The chart shows that, with the exception of one contraction, the two sets of cumulative

[21] Broadly defined, "diffusion" indexes for a group of time series consist of the percentage of these series which are increasing over a specified time span: they measure the degree to which the increases are diffused among the components. In "historical" diffusion indexes, all changes between the troughs and peaks of the component series are regarded as increases, all changes between peaks and troughs as decreases. This means that historical diffusion indexes describe the degree to which cyclical expansion phases prevail among components. For "cumulative" diffusion indexes the differences between the percentage of increasing series and 50 per cent are cumulated, on the theory that these differences reflect the degree of concomitance of upward movements and thus of upward thrust in the group as a whole. If the component series can be aggregated, turns of the simple diffusion indexes lead the corresponding turns of the aggregate and those of cumulative diffusing indexes tend to coincide with them. A basic discussion of the construction and behavior of diffusion indexes can be found in Chapters 2 and 8 of *Business Cycle Indicators,* Geoffrey H. Moore, ed., New York, Princeton University Press for NBER, 1961.

CHART 5

CUMULATIVE HISTORICAL DIFFUSION INDEXES BASED ON PROGRAM-SELECTED AND STAFF-SELECTED CYCLICAL TURNS IN THE COMPONENTS, 1948–65

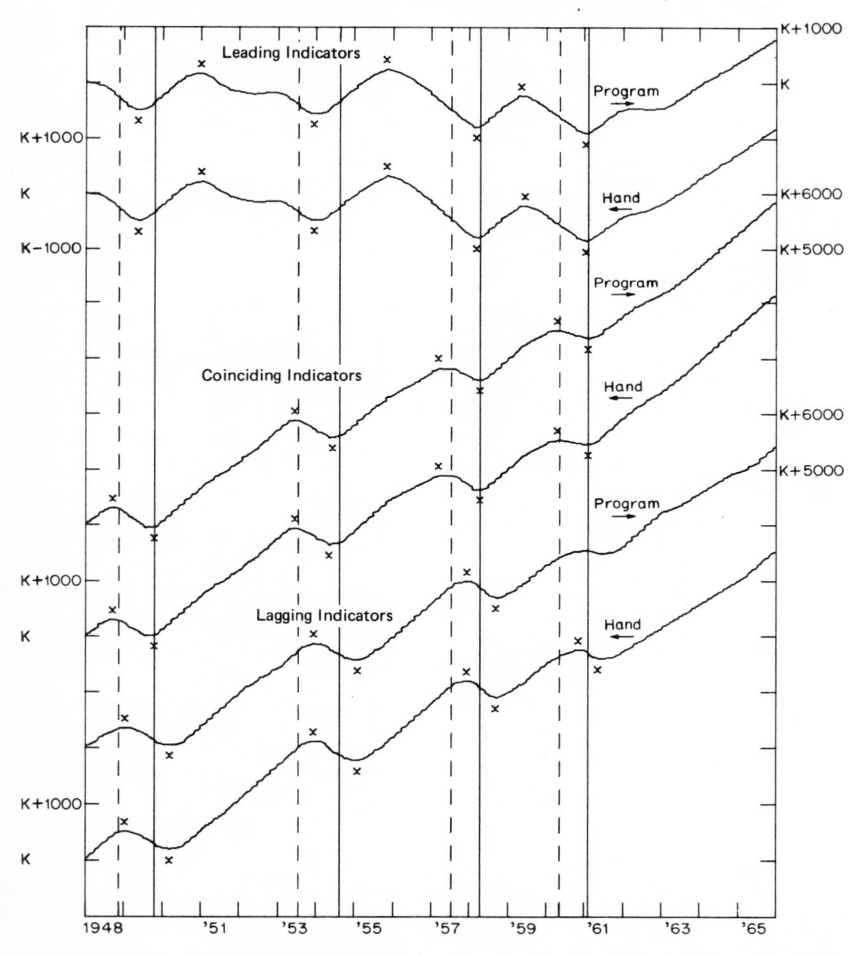

Note: Broken vertical lines denote business cycle peaks; solid vertical lines denote business cycle troughs. Origin of vertical scale is arbitrary, since K may be any constant.

TABLE 5

TURNING POINTS IN CUMULATIVE HISTORICAL DIFFUSION INDEXES, BASED ON PROGRAM-SELECTED AND STAFF-SELECTED CYCLICAL TURNS OF COMPONENTS, 1948–65

Type of Turn	Business Cycle Dates	29 Leading Indicators		15 Roughly Coincident Indicators		7 Lagging Indicators	
		Staff-Selected	*Program-Selected*	*Staff-Selected*	*Program-Selected*	*Staff-Selected*	*Program-Selected*
P	Nov. 1948	—	—	Sept. 1948	Sept. 1948	Jan. 1949	Jan. 1949
T	Oct. 1949	May 1949	May 1949	Oct. 1949	Oct. 1949	Mar. 1950	Mar. 1950
P	July 1953	Jan. 1951	Jan. 1951	June 1953	June 1953	Dec. 1953	Dec. 1953
T	Aug. 1954	Dec. 1953	Dec. 1953	May 1954 *	June 1954 *	Feb. 1955	Feb. 1955
P	July 1957	Nov. 1955	Nov. 1955	Mar. 1957	Mar. 1957	Dec. 1957	Dec. 1957
T	Apr. 1958	Mar. 1958	Mar. 1958	Apr. 1958	Apr. 1958	Sept. 1958	Sept. 1958
P	May 1960	June 1959 *	May 1959 *	Apr. 1960	Apr. 1960	Nov. 1960	—
T	Feb. 1961	Jan. 1961	Jan. 1961	Feb. 1961	Feb. 1961	May 1961	—
LEADS (−), COINCIDENCES (0), AND LAGS (+), IN MONTHS, RELATIVE TO BUSINESS CYCLE TURNS							
P	Nov. 1948	—	—	−2	−2	+2	+2
T	Oct. 1949	−5	−5	0	0	+5	+5
P	July 1953	−30	−30	−1	−1	+5	+5
T	Aug. 1954	−8	−8	−3	−2	+6	+6
P	July 1957	−20	−20	−4	−4	+5	+5
T	Apr. 1958	−1	−1	0	0	+5	+5
P	May 1960	−11	−12	−1	−1	+6	—
T	Feb. 1961	−1	−1	0	0	+3	—

Note: For an explanation of the nature of cumulative historical diffusion indexes, see footnote 21. Asterisks denote a difference between the turning point dates chosen by the two methods.

diffusion indexes move closely together, exhibiting corresponding swings, similar amplitudes, and practically simultaneous timing. The one exception is the 1960–61 contraction in the lagging indicators, which is shown in the index based on staff-selected turns but is omitted in the index based on program-selected turns. Reference to Chart 4 shows that the difference is entirely due to turning point determination in two series (48 and 49). In these series the program does not select turns in 1960 and 1961. The declines during these years are minute and scarcely detectable on our chart.

Table 5 quantifies the relationship of cyclical turns in the two sets of indexes to those in general business activity. The upper panel contains the dates of these turns: only two out of twenty-one comparable turns occurred at different dates (see asterisks), and the differences never exceeded one month. The lower panel focuses on the timing relation of the diffusion index turns to those in business cycles. Note the consistency of signs for both sets of leading and of lagging indexes; most important, note the "practical coincidence" of the turns of cumulative diffusion indexes of coinciding indicators, whether computer based or not, with business cycle turns. That is, with one exception the peaks and troughs of the index either coincide exactly (three out of eight) or occur within three months of business cycle turns. The exception, a lead of four months, occurs at the July 1957 peak. This performance bolsters the hope that program-based cumulative diffusion indexes of coinciding indicators—though not necessarily of the fifteen indicators used above—may play an increasing, and perhaps decisive, role in the determination of business cycle turning points.

One question is whether it is possible to dispense with other evidence, such as amplitudes of cyclical swings and the economic importance of the activities reaching turns at particular dates. Also, the programmed method lends itself better to the identification of past business cycle turns than to the identification of current turns, since it requires evidence for four or more months after the occurrence of a cyclical turn in the component series. Nonprogrammed identification of current business cycle turns may possibly be more prompt. It can more readily make use of other evidence, such as the behavior of leading indicators, the sharpness of the turns, the character and comprehensiveness of specific activities, and the effect of impending events and policies.

There are other applications. The program permits, for example, turning point determination and construction of historical diffusion indexes for large groups of time series, such as sales and profits data of individual companies, or indicators of various economic activities in each of the fifty states. The program can also be used to measure cyclical divisions in levels and changes for business cycle analysis, trend analysis, and other purposes. In short, programmed determination of turning points opens the way for a variety of imaginative experiments.

SUGGESTIONS FOR FURTHER DEVELOPMENT

Experimentation with the programmed turning point selection raises certain issues that should be considered in the further development of the approach.

1. The program is sensitive to the accuracy with which the basic data are reported, that is, rounded data may yield fewer and different turns than do unrounded data. It should be feasible to standardize the input so that the number of digits used does not affect the number and location of specific cycles.

2. The twelve-month moving average, used as the basis for the determination of the presence of cycles, may obliterate shallow but cyclically significant phases in certain series. Conversely, it may transform irregular movements into cyclical patterns. If these tendencies are to be avoided, the smoothing term of the basic trend-cycle representation should bear some relation to the relative importance of irregular, as compared to systematic, movements.

3. The Spencer curve, with its graduated 15-point weight pattern (and negative weights at the ends) is not necessarily the most effective tool for present purposes. It may be fruitful to experiment with other weighting systems and perhaps with flexible weights.

4. The turning points near the ends of the series are frequently more uncertain than others. Some modifications of the approach may be considered. The span (now six months from each end) within which turns are not recognized could be extended. This type of constraint, which now operates only on the turning points of the final (unsmoothed) series, could also be imposed on some of the smoothed curves. Also, the present sequence of tests (first for acceptability of end values, then for cycle and phase durations) might be reversed, so that those turns that become first or last turns only through the re-

jection of end turns would not be subjected to tests designed for turns close to the ends of the series.

5. Present search ranges sometimes exclude values worthy of consideration as cyclical peaks or troughs. While extension of the ranges increases the danger of selecting noncyclical extremes as turns, there is no assurance that the present ranges are optimal. Experiments with alternative ranges might be desirable.

6. Since amplitude considerations are only implicit in the present procedure and in some of the above suggestions, the amplitude effects of these procedures and of contemplated changes should be kept under close review. An explicit amplitude criterion might also be devised and tested.

7. In view of the bias inherent in the use of a collection of well-conforming indicators as a test sample, it is desirable to test the approach on series that present special problems of turning point determination. This will improve our judgment regarding the effectiveness of the approach for economic activities at large.

It is not proposed that all of these suggestions should be evaluated on the basis of present experience. The effects of any specific change, and the interaction of several changes, are hard to foresee. What is needed is some well-organized accounting of the results of the present procedure after it has been more widely used and an evaluation of these results, based on well-defined criteria. When substantial additional knowledge has accumulated, major changes in the current approach might be contemplated.

APPENDIX TO CHAPTER 2

SAMPLE RUN, SELECTION OF CYCLICAL TURNING POINTS, BITUMINOUS COAL PRODUCTION

BITUMINOUS COAL PRODUCTION
100,000 NET TONS 1914 - 1938 01118

NBER TURNING POINT DETERMINATION FOR SERIES 01118

NUMBER OF MONTHS = 300

FIRST MONTH = 1914 1

MCD = 0 CONTROL LIMIT = 3.500

Output Table 2-1

BITUMINOUS COAL PRODUCTION
100,000 NET TONS 1914 - 1938

TIME SERIES DATA 01118

YEAR	JAN	FEB	MAR	APR	MAY	JUNE	JULY	AUG	SEPT	OCT	NOV	DEC
1914	383.0	359.0	382.0	347.0	340.0	345.0	361.0	363.0	371.0	325.0	312.0	336.
1915	351.0	329.0	324.0	349.0	325.0	347.0	356.0	359.0	387.0	384.0	430.0	463.
1916	440.0	508.0	447.0	391.0	408.0	385.0	381.0	403.0	397.0	390.0	432.0	445.
1917	453.0	465.0	489.0	487.0	496.0	478.0	463.0	447.0	425.0	420.0	459.0	444.
1918	398.0	492.0	491.0	535.0	531.0	521.0	550.0	520.0	483.0	455.0	422.0	406.
1938	270.0	261.0	252.0	280.0	254.0	262.0	269.0	296.0	317.0	299.0	329.0	329.

Output Table 2-2

SPENCER CURVE, NO SUBSTITUTIONS 01118

YEAR	JAN	FEB	MAR	APR	MAY	JUNE	JULY	AUG	SEPT	OCT	NOV	DE
1914	370.3	367.0	361.9	357.3	354.6	353.6	353.3	351.7	347.3	341.6	336.3	332
1915	331.4	332.0	333.2	334.9	338.1	343.7	352.5	365.7	383.7	405.4	427.9	446
1916	456.9	454.8	442.7	424.6	406.2	393.2	387.9	389.4	396.9	409.0	423.3	439
1917	456.2	470.8	481.5	486.2	483.6	475.1	462.6	449.1	438.1	431.7	431.3	438
1918	452.7	472.1	494.7	515.5	52[illegible].9	534.4	527.9	511.3	487.1	459.0	430.5	405
1938	291.8	274.0	262.4	258.0	260.3	267.4	278.1	290.4	302.5	312.7	319.3	32

MEAN OF RATIOS TO SPENCER CURVE= 99.91 STD.DEV. = 7.853

EXTREME OBSERVATIONS AND THEIR SUBSTITUTES

71	1919	11	183.0	387.4
99	1922	3	490.0	336.8
100	1922	4	201.0	305.1

Output Table 2-3

BITUMINOUS COAL PRODUCTION
100,000 NET TONS 1914 - 1938

12 MONTHS MOVING AVERAGE 01118

EAR	JAN	FEB	MAR	APR	MAY	JUNE	JULY	AUG	SEPT	OCT	NOV	DEC
1914	359.3	359.5	360.0	361.2	357.6	353.4	352.0	349.3	346.8	342.0	342.1	340.9
1915	341.0	340.6	340.3	341.6	346.5	356.4	367.0	374.4	389.3	399.5	403.0	410.0
1916	413.1	415.2	418.9	419.7	420.2	420.4	418.9	420.0	416.4	419.9	427.9	435.2
1917	443.0	449.8	453.5	455.8	458.3	460.5	460.5	455.9	458.1	458.3	462.3	465.2
1918	468.8	476.0	482.1	487.0	489.9	486.8	483.6	483.6	472.7	461.0	448.0	437.3
1938	309.4	300.9	296.2	290.4	286.1	285.4	284.8	286.1	288.6	292.7	294.3	300.1

Output Table 2-4

4 MONTHS MOVING AVERAGE 01118

AR	JAN	FEB	MAR	APR	MAY	JUNE	JULY	AUG	SEPT	OCT	NOV	DEC
914	367.7	367.7	367.7	357.0	353.5	348.2	352.2	360.0	355.0	342.7	336.0	331.0
915	332.0	335.0	338.2	331.7	336.2	344.2	346.7	362.2	371.5	390.0	416.0	429.2
916	460.2	464.5	446.5	438.5	407.7	391.2	394.2	391.5	392.7	405.5	416.0	430.0
917	448.7	463.0	473.5	484.2	487.5	481.0	471.0	453.2	438.7	437.7	437.0	430.2
918	448.2	456.2	479.0	512.2	519.5	534.2	530.5	518.5	502.0	470.0	441.5	420.2
938	301.2	280.0	265.7	261.7	262.0	266.2	270.2	286.0	295.2	310.2		

Output Table 2-5

TENTATIVE TURNING POINTS, 12 MONTHS MOVING AVERAGE
TENTATIVE PEAKS

	OBS	YEAR	MO	VALUE
1	4	1914	4	361.2
2	53	1918	5	489.9
3	79	1920	7	472.7
4	113	1923	5	484.0
5	154	1926	10	496.6
6	187	1929	7	447.5
7	241	1934	1	308.4
8	280	1937	4	383.7

TENTATIVE TROUGHS

	OBS	YEAR	MO	VALUE
1	15	1915	3	340.3
2	33	1916	9	416.4
3	67	1919	7	403.0
4	99	1922	3	315.1
5	130	1924	10	390.3
6	167	1927	11	399.9
7	227	1932	11	247.4
8	244	1934	4	296.5
9	256	1935	4	300.6
10	295	1938	7	284.8

REJECTIONS

	PEAKS	TROUGHS
MULTIPLE PEAKS OR TROUGHS		33 1916 9 416.4
MULTIPLE PEAKS OR TROUGHS		256 1935 4 300.6

Output Table 2-6

BITUMINOUS COAL PRODUCTION
100,000 NET TONS 1914 - 1938

01118

TENTATIVE TURNING POINTS, SPENCER CURVE B

TROUGHS			PEAKS		
			1914	3	362.
1915	1	331.	1918	6	534.
1919	3	370.	1920	10	486.
1922	6	263.	1923	6	543.
1924	6	377.	1927	1	513.
1927	11	381.	1929	7	464.
1932	6	227.	1934	4	321.
1934	9	279.	1937	1	403.
1938	4	258.			

TEST FOR MINIMUM DURATION OF 15 MONTHS

NO EXCLUSIONS

Output Table 2-7

TURNING POINTS, 4 MONTHS MOVING AVERAGE

TROUGHS			PEAKS		
			1914	3	368.
1914	12	331.	1918	6	534.
1919	3	372.	1920	11	495.
1922	6	230.	1923	6	549.
1924	7	376.	1927	2	518.
1927	12	376.	1929	6	469.
1932	7	218.	1934	4	326.
1934	10	281.	1937	2	418.
1938	4	261.			

Output Table 2-8

BITUMINOUS COAL PRODUCTION
100,000 NET TONS 1914 - 1938 01118

TENTATIVE TURNING POINTS, TIME SERIES

TROUGHS			PEAKS		
			1914	3	382.
1914	11	312.	1918	7	550.
1919	3	350.	1920	12	538.
1922	4	201.	1923	5	568.
1924	6	366.	1927	3	565.
1927	12	370.	1929	5	487.
1932	7	208.	1934	3	363.
1934	9	274.	1937	3	490.
1938	3	252.			

1914 3 382.0 1914 1 383.0 ELIMINATE TURN

TEST FOR MINIMUM DURATION OF 15 MONTHS

NO EXCLUSIONS

FINAL TURNING POINTS, TIME SERIES

TROUGHS			PEAKS		
1914	11	312.	1918	7	550.
1919	3	350.	1920	12	538.
1922	4	201.	1923	5	568.
1924	6	366.	1927	3	565.
1927	12	370.	1929	5	487.
1932	7	208.	1934	3	363.
1934	9	274.	1937	3	490.
1938	3	252.			

3
STANDARD BUSINESS CYCLE ANALYSIS OF ECONOMIC TIME SERIES

GENERAL APPROACH

THE BUSINESS CYCLE ANALYSIS of the National Bureau of Economic Research was originally designed by Wesley C. Mitchell; it was perfected by Arthur F. Burns and Mitchell and is described in detail in their *Measuring Business Cycles,* published by the Bureau in 1946. The following brief description of the method is not, of course, intended as a substitute for that volume, with its wealth of information and discussion. Rather it presents, in skeleton form, the approach and the major measures so that the reader may judge whether they should be considered for his purposes. For basic enlightenment on the analysis, the reader must go back to the original source.[1]

The National Bureau's analysis is designed to be objective, in the sense that different investigators should be able to obtain similar results; it is designed to be stable, in the sense that extension of time coverage should not invalidate the measures previously computed for individual cycles. These two characteristics distinguish the analysis from other approaches that may result in cyclical measures, for any given cycle, that are sensitive to the period selected. The requirement of objectivity facilitates the programming of the analysis for electronic computers. The programmed analysis need not lead to a more rigid procedure since the program contains a great number of options which require exercise of judgment with regard to inclusiveness of analysis, type of measures computed, and variations in the approach selected.

[1] A brief description of parts of the technique may also be found in Wesley C. Mitchell, *What Happens During Business Cycles,* New York, NBER, 1951, pp. 9–25.

The analysis was designed, before the computer age, as a standard method for mass application—that is, a statistical technique whose application could be delegated to persons who did not necessarily have any substantive knowledge of the activity analyzed. The application of the approach, furthermore, led to a set of decision rules for a variety of special circumstances. These features of the analysis also facilitated programming.

It is important to emphasize again that this approach is mainly a scheme to describe, summarize, and compare historical behavior. It was not primarily designed as a method of analyzing current business conditions—although the general results can make distinct contributions toward that end. Analysis of levels or changes during a given period cannot be carried through before the cyclical phase (expansion or contraction) of which that period is a part has been recognized as completed. This involves determination of a terminal turning point, which can be identified only after some time has elapsed. Students whose interest centers around the speedy evaluation of current business conditions must turn to different approaches, such as the recession and recovery analysis described later in this study.

TREATMENT OF TIME SERIES COMPONENTS

In terms of the familiar decomposition of the complete time series into seasonal, trend, cyclical, and irregular components, the National Bureau's measures refer only to the cycle and trend elements. The seasonal component is estimated and then eliminated before a series is subjected to the standard business cycle analysis. This seasonal adjustment process is now, in most cases, carried out by electronic computer. The effect of the random component (minor irregular factors) on the cyclical measures is minimized or at least reduced by various forms of averaging. The effect of episodic events (major irregular factors) is not segregated.

An important characteristic of the standard business cycle analysis is that its cyclical measures include intracyclical trend forces. A separation of cyclical and trend forces is difficult, both conceptually and operationally. It can of course be argued that a rough approximation of such separation is better than none, and that for some purposes—such as demonstrating cyclical characteristics even in rapidly expanding activities—one may wish to observe the cyclical elements with growth excluded.[2] But for many purposes, and particularly when we are concerned with the interrelations among various activities and with their contribution to general business activity, we may want to deal with the "cycle of experience" (albeit after adjustment for seasonal variations). For the maintenance of GNP levels during a contraction, for example, it is more important that state and local government expenditures show strong growth and only small fluctuations in rates of growth than that they show actual declines after removal of trend. General cyclical volatility, on the other hand, is affected by the failure of an activity to maintain its rate of expansion, whether or not this leads to actual declines. Differential growth during expansions and contractions can be measured without previous adjustment for long-term trends. In any case, the basic technique of business cycle analysis can be applied to either trend-adjusted or unadjusted data. Intercycle trends, that is, long-term changes between cycle averages, are described by the secular measures which form part of the business cycle analysis. These will be examined later in this paper.

Erratic movements, such as those occasioned by strikes and natural disasters, as well as randomly distributed smaller irregular movements, may complicate cyclical analysis. In fact, even the first step toward business cycle analysis, the determination of cyclical peaks and troughs in a given activity, may be beset by the problem of irregular highs and lows. Basically, the National Bureau approach tries to deal with the problem of irregular movements by averaging and, in the case of turning point determination, by sometimes disregarding erratic observations. As will be seen below, measures based on levels for a single month are almost never used in the analysis.

SPECIFIC CYCLE AND REFERENCE CYCLE ANALYSIS

The business cycle analysis is carried out within two different frameworks—a specific cycle and a reference cycle framework. In *specific cycle analysis,* cyclical swings of an economic activity are analyzed against a chronological framework that is marked by the upper and

[2] To some extent this can be achieved by analyzing first differences or rates of change. Indications of intracycle trend are also given by the difference between initial and terminal standings in cycle patterns (see pp. 73 ff.) and by the difference between amplitudes during expansions and contractions (pp. 87 ff.). Furthermore, recognition is given to intracycle trend in the computation of full-cycle conformity measures (pp. 107 ff.).

lower turns of the activity itself. This means that the duration or the amplitude of a specific decline is measured between a peak and the succeeding trough of the specific activity, and measurements of full cycles and their subdivisions are based on the same framework. *Reference cycle analysis,* in contrast, measures the behavior of a given economic activity against a chronological framework marked by the peaks and troughs of a reference activity—usually that of business activity at large. (Reference cycle analysis based on other chronologies will be discussed later.) The somewhat ambiguous term—reference cycle amplitude—does not denote the amplitude of fluctuations in the reference activity, but the fluctuations of a specified activity during reference expansions and contractions, that is, between peaks and troughs of the reference activity.

Figure 1 illustrates that the amplitude of a reference cycle contraction (P_r to T_r) cannot, as a rule, be larger than the amplitude of a specific cycle contraction (P_s to T_s) that corresponds to the same business cycle phase, and the same holds for expansions. This is because the specific amplitudes are measured between the extreme highs and lows of the series and can be equal to reference amplitudes only if the peaks and troughs of the specific activity and those of general business conditions coincide. The results of the specific cycle analysis

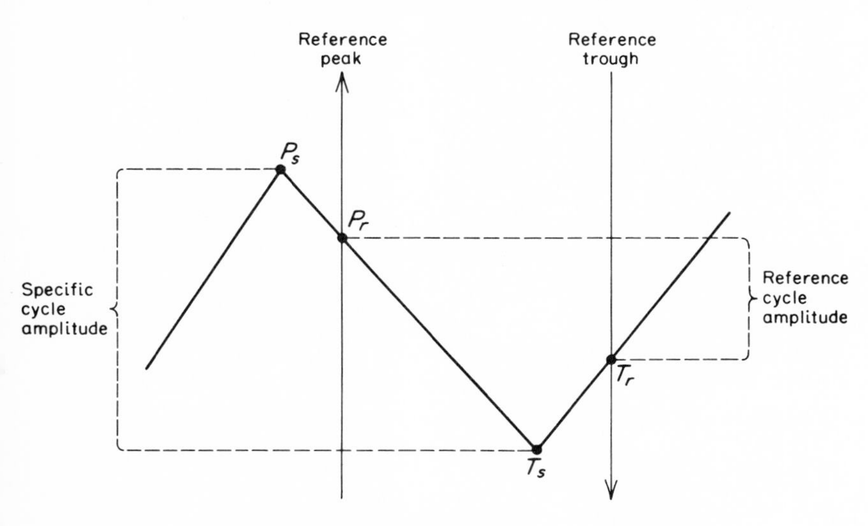

FIGURE 1

and those of the reference cycle analysis are given in two separate sets of tables.

The National Bureau's reference cycle chronology serves as the framework for reference cycle analyses of individual activities, apart from its function of specifying when, in the Bureau's best judgment, general business conditions experienced cyclical peaks and troughs. These reference cycle peaks and troughs are listed, on a monthly, quarterly, and annual basis, in Table 6. They are determined by a rather complex and not very rigidly defined process. In addition to various measures of output (price-deflated GNP, industrial production, and so forth), other activities are considered, such as inputs (of labor and capital), underemployment of resources (human and other), price behavior (in various markets), and monetary phenomena (volume of transactions by check). Some aspects of this determination are discussed in an exchange of views between George W. Cloos and Victor Zarnowitz.[3] For purposes of the analysis of time series against a reference cycle framework, small differences in the business cycle chronology are of little consequence and large changes have rarely been suggested. The most consequential decisions in the establishment of a chronology relate not so much to the precise dating of a particular turning point but to the recognition of a given fluctuation in economic activity as a business cycle. The National Bureau's decision not to recognize the brief decline in many economic activities about a year after the onset of the Korean War as a business cycle contraction is an example. In this case many of the more volatile economic activities registered cyclical contractions, but most of the large aggregates did not. Recognition of that episode as a business cycle contraction would have affected many cyclical measures. For some research purposes it is valuable to include fluctuations such as these in the analysis. In such cases it may be advisable to use reference dates from the subcycle chronology developed by Ruth P. Mack or the growth-cycle chronology developed by Ilse Mintz, both of which contain turning points for short-term fluctuations.[4]

The use of a reference cycle framework for the analysis of economic time series constitutes one way of making comparisons among different

[3] *Journal of Business,* January, April, and July 1963.

[4] See, for instance, NBER, *38th Annual Report,* New York, 1958, p. 31. For growth cycles, see "Dating American Growth Cycles," in *The Business Cycle Today* (forthcoming).

TABLE 6

DATES OF PEAKS AND TROUGHS OF BUSINESS CYCLES IN THE UNITED STATES, 1854–1961

Monthly		*Quarterly*		*Calendar Year*		*Fiscal Year Ending June 30*	
Trough	*Peak*	*Trough*	*Peak*	*Trough*	*Peak*	*Trough*	*Peak*
				1834	1836		
				1838	1839		
				1843	1845		
				1846	1847		
				1848	1853		
December 1854	June 1857	IV 1854	II 1857	1855	1856		
December 1858	October 1860	IV 1858	III 1860	1858	1860		
June 1861	April 1865	III 1861	I 1865	1861	1864		
December 1867	June 1869	I 1868	II 1869	1867	1869	1868	1869
December 1870	October 1873	IV 1870	III 1873	1870	1873	1871	1873
March 1879	March 1882	I 1879	I 1882	1878	1882	1878	1882
May 1885	March 1887	II 1885	II 1887	1885	1887	1885	1887
April 1888	July 1890	I 1888	III 1890	1888	1890	1888	1890
May 1891	January 1893	II 1891	I 1893	1891	1892	1891	1893
June 1894	December 1895	II 1894	IV 1895	1894	1895	1894	1896
June 1897	June 1899	II 1897	III 1899	1896	1899	1897	1900
December 1900	September 1902	IV 1900	IV 1902	1900	1903	1901	1903
August 1904	May 1907	III 1904	II 1907	1904	1907	1904	1907
June 1908	January 1910	II 1908	I 1910	1908	1910	1908	1910
January 1912	January 1913	IV 1911	I 1913	1911	1913	1911	1913
December 1914	August 1918	IV 1914	III 1918	1914	1918	1915	1918
March 1919[a]	January 1920	I 1919[a]	I 1920	1919	1920	1919	1920
July 1921[a]	May 1923	III 1921	II 1923	1921	1923	1922	1923
July 1924	October 1926	III 1924	III 1926	1924	1926	1924	1927
November 1927[a]	August 1929[a]	IV 1927	III 1929[a]	1927	1929	1928	1929
March 1933	May 1937	I 1933	II 1937	1932	1937	1933	1937
June 1938[a]	February 1945	II 1938	I 1945	1938	1944	1939	1945
October 1945	November 1948	IV 1945	IV 1948	1946	1948	1946	1948
October 1949	July 1953	IV 1949	II 1953	1949	1953	1950	1953
August 1954	July 1957	III 1954	III 1957	1954	1957	1954	1957
April 1958	May 1960	II 1958	II 1960	1958	1960	1958	1960
February 1961		I 1961		1961		1961	

Source: Burns and Mitchell, *Measuring Business Cycles,* Table 16, except for revisions noted and dates since 1938.
[a] Revised.

activities. If the length of the average workweek leads a reference turn by four months and employment leads by one month, obviously the workweek leads employment by three months. Cyclical patterns and amplitudes of two activities for identical periods can also be compared on the basis of the measures derived by the Bureau's reference cycle analysis. However, on occasion one might want to know how the workweek changed as employment declined from employment peaks to employment troughs, how prices changed while inventories accumulated, or perhaps how inventories changed during periods of cyclical price declines. Here the period of comparison would be based on the turning points of a single series. It is possible to make such comparisons by using the peaks and troughs of the comparative series (such as employment in our first example) as a reference framework. Technically, this is done by simply substituting the specific turning points of the reference series for the turns in general business conditions.[5]

SUBDIVISION OF CYCLES

Expansions and Contractions, TPT and PTP Cycles. The elementary division of cycles (business cycles or specific cycles) into two phases—expansions and contractions—has been mentioned. Historically, the National Bureau analysis combined these phases into trough-peak-trough (TPT) cycles, that is, into cycles encompassing an expansion and the subsequent contraction. Mitchell and his associates were always aware of the fact that contractions breed recoveries no less than expansions breed recessions, so that peak-trough-peak (PTP) cycles would form equally acceptable—albeit equally artificially delineated—units of measurement and analysis. Only conservation of time and effort prevented this consideration from being reflected in a generalized analysis on both bases, TPT and PTP. The business cycle computer program now provides both types of analysis optionally. This, incidentally, has the practical advantage that a complete recent cycle can also be analyzed when the last-experienced phase is an expansion. Furthermore, the use of both analyses, that is, analyses for overlapping TPT and PTP cycles, may help to overcome the artificial segmentation of dynamic events into separate cycles with somewhat restricted comparability. Now every phase can be compared, on a common cycle base, with the contiguous preceding and subsequent phases.

Indeed, in many economic series the amplitude of given expansions is more highly correlated with preceding than with subsequent con-

[5] For an example of this use of a reference framework, see Thor Hultgren, *Cost, Prices, and Profits: Their Cyclical Relations,* New York, NBER, 1965, especially pp. 56–57.

tractions. "What goes down will come up" seems to be more regularly true in economic activities than "what goes up must come down." This suggests that the peak-to-peak cycle may be a more homogeneous unit than the trough-to-trough cycle, and upon reflection this appears plausible. The degree of underutilization of resources, in a rough way, measures both the opportunity for and the short-term limits of expansion. After a deep recession, recovery to previous peak levels is expected, at the least, as well as some making up for postponed investment in producer and consumer goods and some further growth in response to population, labor force, and productivity increases. The situation is not analogous during recessions. High peak levels—except if accompanied by highly speculative booms in construction, overbuilding of equipment and inventories, and blatant overvaluation of assets —may not tell much about the severity of the following contraction. However, the matter is further complicated when short and mild contractions but long and vigorous expansions are experienced. Here the recovery phase—that is, the period during which economic activities regain former peak levels—covers a relatively small portion of the total expansion, and the recovery amplitude forms a small portion of the total upswing. This tends to loosen the relation between contractions and subsequent expansions. Chart 6, showing the fluctuations of the Federal Reserve Board Index of Industrial Production during the interwar and the postwar periods, illustrates this point. During the interwar period, the recovery segment tended to account for a large part of the full expansion. This is related to the occurrence of severe contractions as well as to the relative weakness of growth. By contrast, the postwar expansions are strongly affected, even dominated, by the growth segments. This is particularly the case for the long expansions of 1949–53 and the one beginning in 1961.

It is, of course, possible to regard each cyclical expansion phase and each cyclical contraction phase as a basic unit of measurement and analysis. This has been done in the past, particularly for amplitude comparisons.[6] One advantage of this procedure is that the amplitude

[6] In this case the phase average could serve as the base for the computation of relative amplitudes (peak minus trough, divided by phase average). As a shortcut, the average between the peak and trough standings has frequently been taken instead of the full-phase average. For a discussion of this measure, see Julius Shiskin, *Signals of Recession and Recovery,* New York, NBER, 1961, p. 123. For an application, see Gerhard Bry and Charlotte Boschan, *Economic Indicators for New Jersey,* New Jersey Department of Labor and Industry,

CHART 6

RECOVERY SEGMENTS AND
GROWTH SEGMENTS OF EXPANSIONS,
FEDERAL RESERVE INDEX OF INDUSTRIAL PRODUCTION,
1920–39 AND 1946–65

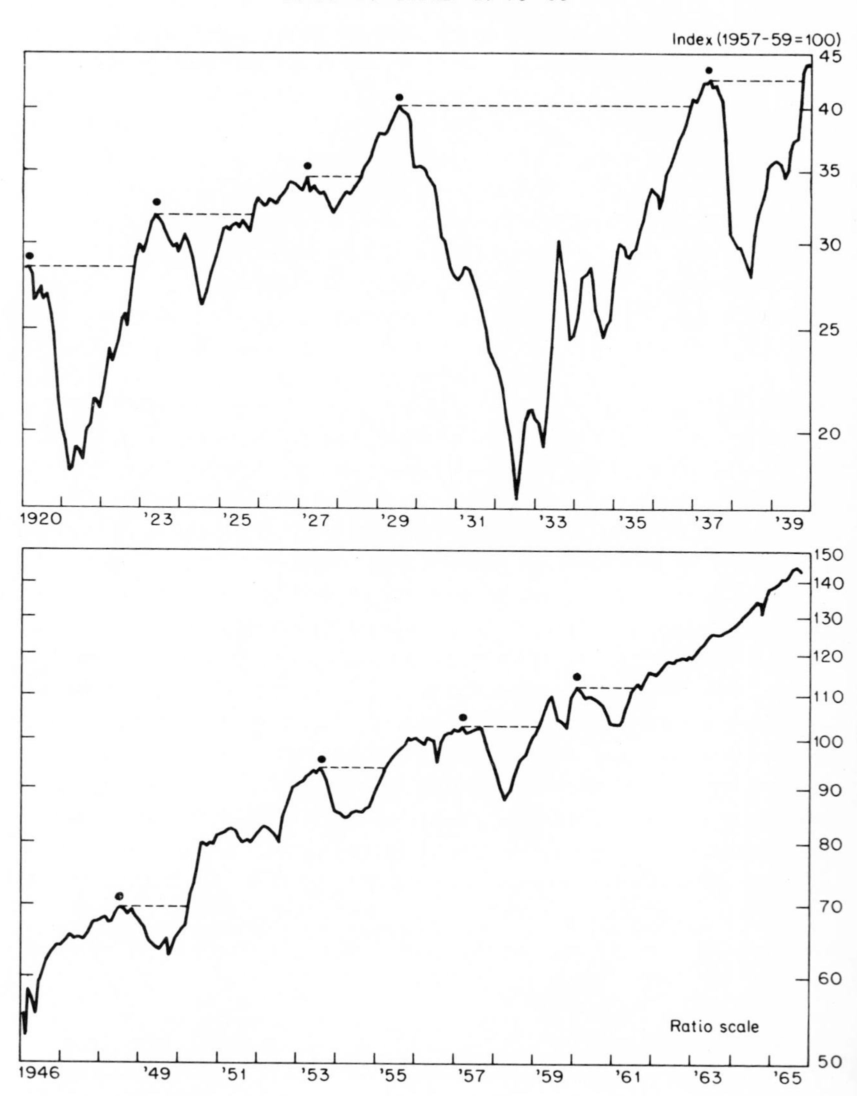

Note: Adjusted for seasonal variations.

CHART 7
REFERENCE CYCLE PATTERNS
OF SIMILAR AMPLITUDE

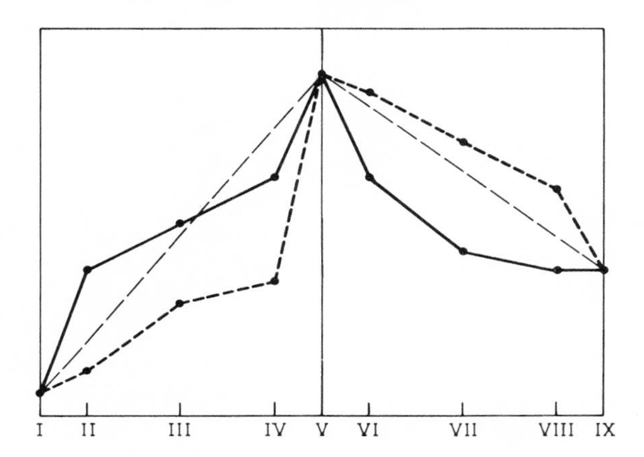

measure for one phase is not affected by the average standing of contiguous phases. Another is that trend measures, which are sometimes computed on the basis of overlapping full-cycle averages (see p. 111, below), could be simplified by basing them on successive phase averages. A disadvantage, however, is that contiguous phase-amplitudes do not have a common base.

Cycle Stages. The division of cycles into two phases permits the computation of durations and amplitudes for expansions, contractions, and full cycles, but it gives little information on intermediate movements. Any detailed analysis of cyclical movements should certainly distinguish behavior such as that illustrated by the two curves of Chart 7—curves that differ, although they may have the same duration and phase amplitude.

The approach of Burns and Mitchell is to divide the cycle—both reference and specific—into nine stages. The three-month averages centered at the initial trough, the subsequent peak, and the terminal trough are termed stages I, V, and IX, respectively. The intervals between stages I and V are divided (as equally as possible) into three

Division of Employment Security, 1964, p. 23. Currently, Milton Friedman and Anna J. Schwartz are using phase averages instead of cycle averages as bases for business cycle and trend analysis. Tables containing these measures are incorporated into the program on an optional basis.

parts, and the average standing for each of these parts is the standing of stages II, III, and IV, respectively. Similarly, the intermediate thirds between stages V and IX provide the standings for stages VI, VII, and VIII. The intervals between the midpoints of I and II, IV and V, V and VI, and VII and IX are smaller than those between the other adjoining stages. All this can be clearly seen in Chart 7.

Phase Fractions Versus Chronological Measures. An important aspect of the approach just described is that the division of phases into stages is based on fractions of the phase rather than on fixed chronological time spans. That is, phase fractions rather than months form the bases of the stage standings. It is implicitly assumed that the early thirds, the middle thirds, and the late thirds of expansions (or contractions) have more in common than, say, the first six months, the middle six months, and the last six months of different cycles. It is possible that generalizations concerning, say, the characteristics of the late periods of expansion may be more nearly valid when expressed in chronological units (a fixed number of months) than in fractions of a phase (which may last as little as two months for a short phase and a year or more for a long one). Whether this is so or not may change from activity to activity or from cycle to cycle. It could be made a subject of research (1) if comparisons are made between the patterns derived by the National Bureau's recession and recovery analysis (which is carried out in units of chronological time) and those of the standard business cycle analysis; or (2) if leads and lags measured in months were compared with leads and lags measured as a fraction of the cycle phase. In any event, an advantage of the present procedure is that each cycle has the same number of stages (nine), which is convenient for comparing cycles and averaging the patterns of different cycles.

RELATIVES AND AVERAGES

Cycle Relatives. The average standings for each stage of the nine-stage patterns are first expressed in the original units of the series subjected to analysis. These units may be short tons, dollars, percentages, utilization rates, or any others. Occasionally one may wish to use information in this form. If, let us say, the components of an aggregate such as GNP or the dollar sales of different products are being analyzed, it may be desirable to make comparisons in original units, which can be added up. Or, if different interest-rate series are being compared, the relevant comparison may be in terms of the original units, as when differentials in terms of "basis points" are derived. On the other hand, the diversity of units in which economic series are expressed has severe disadvantages if we attempt to compare cyclical behavior in different cycles, different activities, or different regions. It would be difficult indeed to state whether employment or weekly earnings have larger intracyclical movements as long as their cycle standings are expressed in different units.

This problem might be attacked by computing percentage changes or perhaps "relatives" that describe amplitudes independently of the original units of measurement. Here another difficulty must be faced—that of the changing base in terms of which the percentage or the relative is expressed. A price change from 60 to 90 cents per pound would be expressed as a 50 per cent increase, while the same absolute change from 90 to 60 cents would be only a 33⅓ per cent decrease. Thus, in percentage terms as well as in terms of relatives, the same change in absolute units would seem to loom larger if measured as expansion than if measured as contraction.

Both problems—differing units and percentage-base bias—can be solved by expressing the original values as percentages of a common base. The National Bureau's business cycle analysis takes the average for any full cycle as a base and expresses the cycle standings as relatives with respect to this base, that is, the cycle standings are converted into cycle relatives (Output Table 3A-4).[7] The cycle averages can be computed for peak-to-peak or for trough-to-trough cycles.[8] At this point, the choice of one or the other method may lead to differences in results. Thus, if resources permit, the analysis should be performed on both bases. Output Table 3A-3 contains stage standings in their original units, Output Table 3A-4 contains cycle relatives on a trough-to-trough basis, and Output Table 3A-11 has cycle relatives on a peak-to-peak basis.

[7] Output tables are the end result of electronic computer programs. The output table numbers cited in this paper refer to the output tables reproduced in the appendixes. A given output table number may not always refer to the same part of the analysis, depending on the option chosen. (The program provides for consecutive numbers irrespective of table content.) In the following analysis, the tables will be referred to without regard to their sequence, since this is largely determined by purely computational requirements.

[8] In order to avoid overweighting of peaks in peak-to-peak cycles and of troughs in trough-to-trough cycles, the first and last months of each cycle are given only a half weight in the computation of cycle averages. This also ensures that the individual cyclical standings are compatible with the original data and their over-all average.

The conversion to cycle relatives makes it possible, for instance, to derive amplitude measures for expansions and contractions by simply computing the differences between the cycle relatives at peaks and at troughs. In nonagricultural employment, for example, the amplitude of the 1933–37 expansion is thus +31.2 (peak, 112.3, minus preceding trough, 81.1) and that of the subsequent contraction 1937–38 is −11.2 (trough, 101.1, minus preceding peak, 112.3), both expressed as a percentage of the average standing during the cycle running from 1933 to 1938. The full-cycle amplitude is 42.4, obtained by subtracting the amplitude of contraction from that of expansion. These amplitude measures will be found in Output Table 3A-2.

The magnitude of changes between different stages can also be derived by the same process. Thus in Output Table 3A-5 the rise between stages II and III of the same cycle is measured as 6.9 (i.e., 97.1 minus 90.2, as given in Output Table 3A-4). Ordinarily the rate of change per month is of more interest than the total change between given cycle stages. To obtain such measures, the changes are put on a monthly basis by simply dividing them by the number of months covered between the midpoints of the stages compared. The resultant monthly amplitudes are part of the regular program output (see, for instance, Output Tables 3A-6 and 3A-13). They indicate, for example, whether employment increases slow down as an expansion proceeds.

Averaging of Cycles. The subdivision of each cycle into an equal number of phases and stages and the expression of the standings (average levels) during these stages in terms of cycle relatives provide a basis for summarizing and averaging. If cycle stages are regarded as functionally comparable (standardization of time relative to cycle phases), and if standings are regarded as comparable when expressed as a percentage of cycle averages (standardization of levels relative to cycle averages), summarizing the cyclical behavior of a given activity over several cycles is a meaningful procedure. Chart 8 illustrates the process by presenting cycle patterns for several separate cycles and the average cycle pattern for all the cycles included. The average cycle pattern is of course beset with the problem inherent in all averages based on a relatively small number of observations—the problem of representativeness. Average cycle patterns for the interwar period may be dominated by the strong decline and the correspondingly steep recovery of the Great Depression and its aftermath, and the averaging

CHART 8

REFERENCE CYCLE PATTERNS, NONAGRICULTURAL EMPLOYMENT AND UNEMPLOYMENT RATE, 1933–61

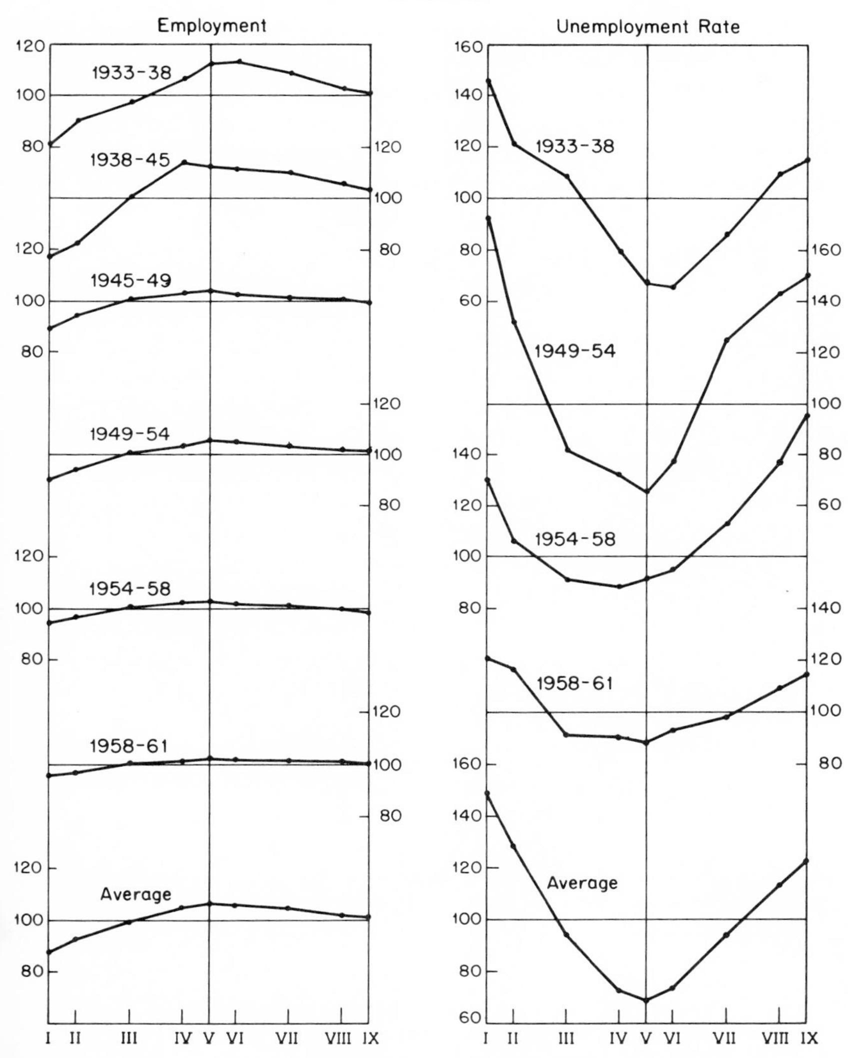

Note: War cycles are not charted separately for the unemployment rate, but are included in the average pattern.

of that cycle pattern and of adjoining milder cycles may produce an average pattern that represents neither the mild nor the strong cycle. In principle it is possible to exclude unusual cycles, such as war cycles, from the averages—or several sets of averages may be computed.

An average is merely a summarizing device, and it is necessary to look not only at the average but also at the dispersion around it and at the behavior of the contributing elements. The output tables of the National Bureau's business cycle analysis always offer all these measures, and the corresponding charts contain both average cycle patterns and those of the individual cycles that comprise the averages.

PROGRAMMED MEASURES OF CYCLICAL CHARACTERISTICS

One way to become acquainted with the results of the National Bureau's business cycle analysis is to go, step by step, through the output tables provided by the computer program and reproduced in the appendixes. This is done here for two sample series describing the number of employees in nonagricultural establishments (Appendix 3A) and unemployment as a percentage of the civilian labor force (Appendix 3B). These series illustrate several versions of business cycle analysis. One version is illustrated by a positively conforming series (nonagricultural employment); the other (unemployment rate) represents an inversely conforming activity. The analysis of the latter, furthermore, is based on absolute changes of rates rather than on changes in cycle relatives. This will be further explained below in the systematic discussion of output tables.

The basic time series are reproduced in Output Tables 3A-1 and 3B-1. With respect to these tables, it should be noted that:

1. All business cycle measures are based on seasonally adjusted data.[9]

2. The number in the upper right-hand corner is an identification number.

[9] The process of seasonal adjustment used can be performed on an electronic computer. See Julius Shiskin and Harry Eisenpress, *Seasonal Adjustments by Electronic Computer Methods,* New York, NBER, Technical Paper 12, 1958; Shiskin, *Electronic Computers and Business Indicators,* New York, NBER, Occasional Paper 57, 1957; and *Tests and Revisions of the Census Methods Seasonal Adjustments,* Washington, D.C., Bureau of the Census, Technical Paper No. 5, 1961.

3. The output is printed without decimal points and may thus appear in unconventional units. (The unemployment rate, for example, is usually stated as a percentage of the civilian labor force but is here printed as per cent × 100.) The units specified are printed under the title.

4. The words "absolute changes" provided after the table heading of the unemployment rate mean that in this analysis measures are computed in terms of differences rather than ratios. This is done because unemployment rates are expressed as a percentage of the civilian labor force—i.e., they are already in relative form and in this respect their levels are comparable over time.

It has been proposed that business cycle analysis should be based on a smoothed rather than unsmoothed version of the seasonally adjusted data. For example, one might analyze a short-term moving average of the seasonally adjusted data (such as the MCD curve of the seasonal adjustment program) or a weighted long-term moving average (such as a Spencer or a Henderson curve). It can be argued in favor of such a procedure that (1) the series, after adjustment for seasonal and irregular movements, constitute conceptually the better approximation to the subject of analysis, that is, cycles and trend; (2) turning point determination is less affected by irregular movements; (3) cyclical characteristics appear more clearly if freed from nonsystematic influences. There are also some considerations, mainly of a practical nature, that militate against the proposal: (1) a uniform smoothing term for a heterogeneous series may not effectively screen out the irregular component; (2) moving averages tend to cut off some of the cyclical amplitude, produce rounded peaks and troughs, and often distort timing relationships; (3) different smoothing terms may produce different cyclical measures; and (4) considerable smoothing is already incorporated in the analysis. For series with highly irregular movements, such as series for individual companies, analysis based on smoothed data may be advantageous. The use of MCD curves would tend to reduce all series to an equivalent degree of smoothness.[10] Irregular series should perhaps be analyzed both in smoothed and unsmoothed form.

[10] For an explanation and illustration of MCD curves, see Shiskin, *Electronic Computers and Business Indicators,* pp. 236–243, or U.S. Department of Commerce, "Description and Procedures," *Business Cycle Developments* (any issue). For the effects of smoothing on cyclical measures, see Arthur F. Burns and Wesley C. Mitchell, *Measuring Business Cycles,* New York, NBER, 1946, Chapter 8.

TIMING AND DURATION OF CYCLES

Timing and duration measures require prior determination of specific turning points (assuming reference dates are given). The process of turning point determination and an experimental computer program for such determination were described in Chapter 2.

Timing Comparisons. Timing measures describe the relation between specific turning points of a time series and corresponding business cycle turning points or other cyclical benchmarks. These measures are convenient, since relating specific turns of many series to common benchmarks makes it possible to compare the timing of turns, for any number of series, without getting entangled in a web of individual comparisons. The study of timing relations for economic activities has many applications. It helps in understanding the mounting and lessening pressures within the economic system in the neighborhood of the crucial periods around business cycle peaks and troughs. Knowing that labor income decreases promptly around business cycle peaks but wage rates decline late (or not at all) permits an understanding of how profit margins are squeezed simultaneously by reduced demand and rigid cost elements. The sequence of turns in economic activities may also reveal something about the causal sequence of economic events. However, great caution must be exercised in the interpretation. The example of maintained wage rate levels illustrates how economic effects may be induced by the very absence of change in a variable. Timing relations are also highly useful for purposes of current business condition analysis and forecasting. Knowledge about the typical timing behavior of economically strategic activities forms the basis of the "indicator approach" to forecasting.[11]

Once specific turns are established, one can determine timing relations for those turns that can be related positively (specific peaks corresponding to reference peaks) or invertedly (specific peaks corresponding to reference troughs) to business cycle turns. Whether a series should be related positively or invertedly can be established either by general considerations, visual inspection, or computation of formal measures. (These so-called conformity measures are discussed later in this section.) This decision can become difficult when the activity conforms only irregularly to business cycles or when it tends to precede

[11] See Geoffrey H. Moore, ed., *Business Cycle Indicators,* Princeton, N.J., Princeton University Press for NBER, 1961, 2 vols.

or to follow general business activity by substantial intervals. In the latter case a problem arises as to whether the series should be regarded as leading on a positive basis or as lagging on an inverted basis, and vice versa.[12] This problem does not exist in the sample series. Employment conforms positively, unemployment rates conform invertedly, and the leads and lags are not large enough to shed doubt on this relationship.

Even after the issue of conformity is settled, the matching of specific turns and reference turns can be problematic. Problems may occur if there is no one-to-one relationship between matching peaks or between matching troughs. Under those circumstances it may be difficult to determine which specific turn should be related to which reference turn. These decisions involve many considerations such as proximity, typical timing behavior at other turns, and relative amplitudes. This is why the timing measures described in the present study are based on visual matching rather than on computerized matching.[13]

Schematic timing charts—which show, for a specific series, leads (−), lags (+), and coincidences (0) in months for each turn—are reproduced in Chart 9. This and the following tabulation permit both a visual and arithmetic evaluation of the relation of specific to reference cycle turns.

	Employees in Nonagricultural Establishments, 1939–61		*Unemployment Rate, 1933–61*	
	Mean	*Median*	*Mean*	*Median*
Peak	−3.3	−1	−3.8	−4
Trough	0.0	0	+2.3	+2

In the case of employees in nonagricultural establishments, 1929–61, troughs coincide, on the average, with business cycle troughs, and

[12] Examples of series which may be regarded as positively lagging or inversely leading are manufacturers' inventories of finished goods, unit labor cost in manufacturing, and bank rates on business loans. For a brief discussion of the economic rationale of treating them either way, see Geoffrey H. Moore and Julius Shiskin, *Indicators of Business Expansions and Contractions,* New York, NBER, Occasional Paper 103, 1967, pp. 30–31.

[13] Since the completion of the present study we have developed an approach to a fully computerized derivation of timing measures. See NBER, *Annual Report,* 1970.

CHART 9

TIMING OF TURNING POINTS RELATIVE TO BUSINESS CYCLE TURNS, NONAGRICULTURAL EMPLOYMENT AND UNEMPLOYMENT RATE, 1929–61

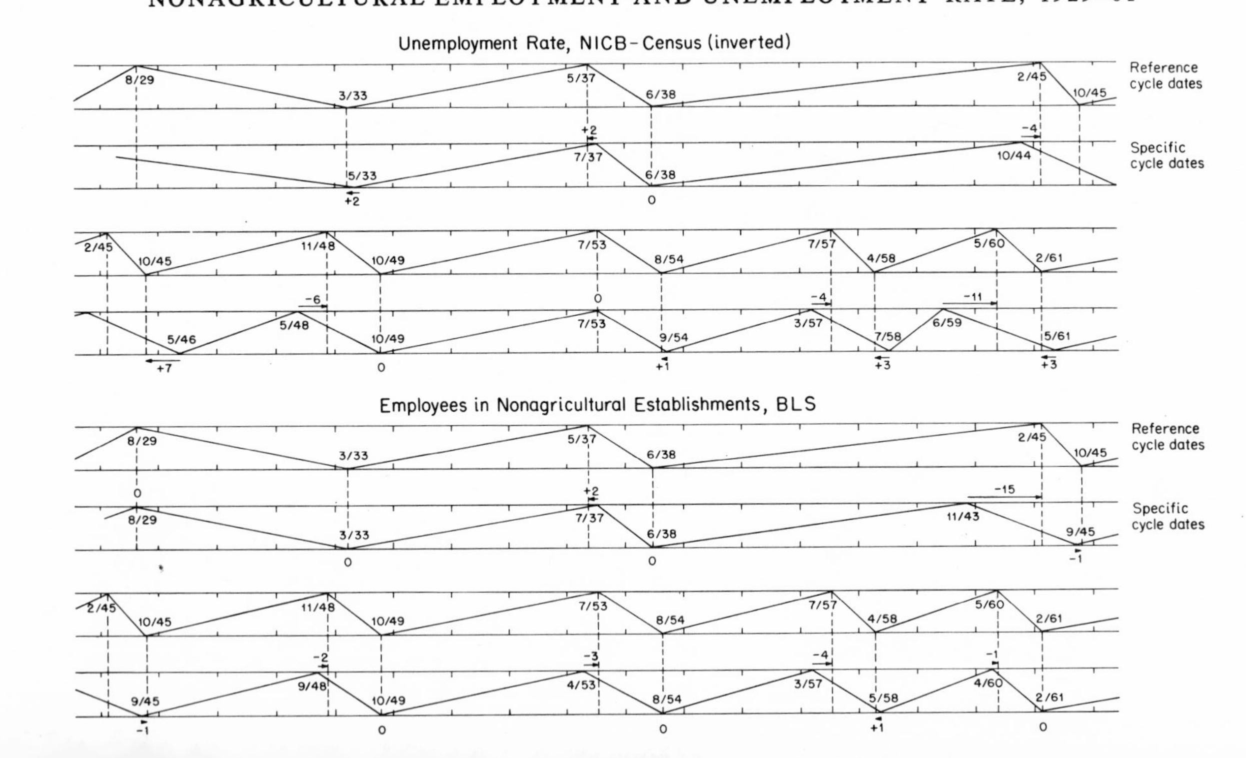

peaks show a mean lead of 3.3 months and a median lead of one month. However, the series can hardly be regarded as typically leading at peaks since the mean is largely affected by a fifteen-month lead during World War II. If that were excluded, the mean lead would shrink to little more than one month. Stronger systematic deviations from reference cycle turns are apparent in the timing charts for the unemployment rate. The tabulation shows substantial mean and median leads at peaks and short lags at troughs. Reference to Chart 9 shows that these average timing characteristics prevail, with few exceptions, throughout the period for which data are available. The averages are not dominated by one or two exceptionally long leads or lags.

The tabulation of the average timing relations of employment and unemployment relative to reference turns permits direct comparisons between the two variables. On the average, unemployment leads employment by 0.5 months (3.8 minus 3.3) at peaks (that is, there is no significant difference), while it lags employment by 2.3 months at troughs. In this sort of comparison between the timing of several variables, care must be taken that the same periods (or rather the same reference turns) are covered. Furthermore, the timing measures will be truly comparable only if there is a one-to-one correspondence between the specific turns compared. In comparing the timing of several activities, additional evidence on timing relations can sometimes be adduced by inclusion of cycles they have in common, even though these cycles do not correspond to business cycles. If such common nonconforming cycles pervade a whole collection of series—as may happen in the analysis of regional data or data for a specific industry or company—it may be preferable to relate the turning points of the component series to the turns of a series representing the cyclical behavior in the particular segment rather than in the economy as a whole (see pp. 68–70).

Some comments are in order on the statistical and economic meaning of these measures. That total nonagricultural employment shows little systematic deviation from business cycle turns is plausible enough. Not only would labor input, as reflected in the number of persons employed, be expected to fluctuate with general business conditions, but the behavior of the employment series is one among several comprehensive measures considered in the dating of business cycle turns. However, the systematic lead of the unemployment rate around business cycle peaks and its lag around troughs require explanation. Such timing behavior might be expected, for statistical reasons, from any inverted series with rising trend—and the unemployment rate does indeed exhibit such a trend during the recent postwar period. However, this trend was relatively mild and has been clearly discernible only since the mid-1950's, whereas the described timing pattern was also discernible before the mid-1950's. Let us therefore look at the timing behavior of the unemployment rate from another point of view.

The number of unemployed is the difference between the number in the civilian labor force and the number employed. The labor force grows relatively steadily, with only mild cyclical variations in its rate of growth. Employment, on the other hand, shows pronounced and relatively smooth cycles. Before the peak in employment, a period generally occurs during which employment rises only slightly while the labor force continues to grow at about the same rate as before. This causes unemployment to increase before employment begins to decline. Expressing unemployment as a rate (relative to the labor force) may cut this lead a little but does not obliterate it. After the trough in employment, there is a roughly analogous situation. The continuing growth of the labor force tends, for a short while, to exceed the employment rise; thus unemployment continues to increase. However, since the characteristic pattern is that of a stronger response of employment to recovery forces than to contracting forces (see Output Tables 3A-22 and 3B-22), the lag of the unemployment rate at business cycle troughs tends to be shorter than its lead at peaks. This explanation, admittedly, does not go much beyond the mechanics of the timing relationship. However, to analyze the causes of the cyclical and growth characteristics of labor force, employment, and unemployment would exceed the scope of this book.

Duration Measures. Reference was made earlier to the connection between timing relations and the duration of expansions and contractions. A well-conforming series without skipped or extra cycles and with roughly coincident timing (or with similar timing at peaks and at troughs) will, of course, show cyclical durations of expansions and contractions close to those of business cycles during the corresponding period. Thus the average duration of cycle phases in employment for the years 1933–61 was rather close to those of business cycles, particularly if the war cycle 1938–45 is excluded. By contrast, the lead of unemployment before business cycle peaks and its lag after troughs makes for longer specific contractions—that is, for a lengthening of the period during which unemployment rises.

The evidence for these observations is summarized in Table 7 and stems from Output Tables 3A-29, 3B-29, 3A-36, and 3B-36 of the electronic computer program for specific cycle durations, and from Output Tables 3A-9, 3B-9, 3A-16, and 3B-16 for reference cycle durations. Care must be exercised that in the comparison of duration and other measures among different activities, comparable time periods are used. The time periods need not be identical, since the specific turns at the beginning and end of the period may differ from activity to activity. Also, the number of specific cycles is not necessarily the same, since an activity may skip cycles or show extra cycles. For certain purposes one might want to compare only corresponding cycles. That summary measures are given in the same computer run and, hence, presumably cover the same over-all time period is no guarantee of the comparability of the cyclical measures. Take, for instance, the duration measures (or for that matter, any other mea-

TABLE 7

DURATION OF EXPANSIONS, CONTRACTIONS, AND FULL CYCLES, NONAGRICULTURAL EMPLOYMENT AND UNEMPLOYMENT RATE, 1933–61

	Average Duration (months)		
	Expansions	*Contractions*	*Full Cycles*
ALL CYCLES			
Employees, nonagricultural establishments	41.7	14.2	55.8
Unemployment rate (inverted)[a]	38.3	17.7	56.0
Business cycles	45.3	10.5	55.8
EXCLUDING WAR CYCLE, 1938–45			
Employees, nonagricultural establishments	37.0	12.6	49.6
Unemployment rate (inverted)[a]	30.8	17.4	48.2
Business cycles	38.4	11.0	49.4

[a] Since unemployment is inversely related to business cycles, the duration measures for expansions, given above, refer to the number of months during which the unemployment rate declines, and those for contractions to the period during which the rate rises.

sures) of the unemployment rate for peak-to-peak cycles as given in Output Table 3B-16 for reference cycles and in Output Table 3B-36 for specific cycles. The summary measures do not refer to corresponding cycles, since the specific analysis omits the 1929–37 cycle because no specific turn could be established in 1929. Note, incidentally, that the whole 1929–37 cycle, and not only the 1929–33 contraction, was omitted. This occurred because the electronic output tables contain duration measures only for full cycles, be they trough-to-trough or peak-to-peak cycles. This means that a cycle phase at the beginning and at the end of any series will be neglected either by the PTP or by the TPT analysis, though not by both.

Chart 10 shows, in schematic fashion, the phases included in the average duration measures of the standard output tables. Note that the duration measures of the PTP analysis include all phases of series starting and ending with peaks, but omit initial expansions of all series starting with troughs, and terminal contractions of all series ending with troughs. The duration measures included in the TPT analyses show analogous inclusions and omissions. The averages provided by

CHART 10

PHASES INCLUDED IN AVERAGES, PEAK-TO-PEAK AND TROUGH-TO-TROUGH ANALYSES

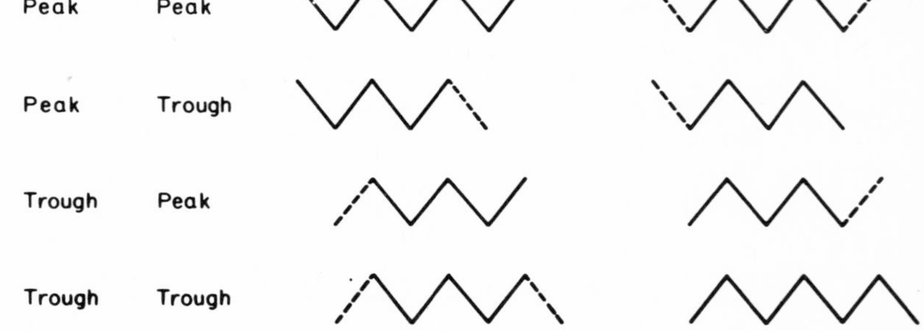

——— Phases included

------- Initial and terminal phases omitted in averages of specified analyses

Series		Type of analysis	
Starts at	Ends at	Peak to peak	Trough to trough
Peak	Peak		
Peak	Trough		
Trough	Peak		
Trough	Trough		

the computer program have the merit of consistency—that is, the average expansion plus the average contraction equals the average full cycle (barring rounding discrepancies). However, if instead it is desired that the averages of phase durations include all available information, this can be easily done on the basis of the schematic drawings in Chart 10. For series beginning and ending at a like turn, the computed averages of one or the other analysis can be used. For series beginning and ending at unlike turns, the average has to be recomputed to include an initial or terminal phase duration. The duration measures for these phases are, of course, provided by the computer run.

The preceding comments have a broader implication that should be made explicit: summary measures, whether electronically computed or otherwise, should not be used without a meticulous check on their composition. The computer output, with its detailed information on intermediate values, component measures, measures derived by alternative approaches, and so forth, facilitates comparisons and interpretations. However, this output must be competently utilized, otherwise the huge output that is generated by programmed analyses may become the massive support of dubious conclusions.

AMPLITUDES DURING EXPANSIONS AND CONTRACTIONS

Value of Amplitude Measures and Variety of Approaches. How large are the cyclical swings in employment and unemployment? Are they growing or declining in size? The general importance of amplitude measures can be readily understood by reference to these variables. The magnitude of cyclical swings in employment are relevant for understanding the socioeconomic aspects of fluctuations in market demand, the labor-input limitations of expansions, the fluctuations in income and expenditures, unemployment changes, and many other phenomena. A similar list could easily be drawn up to illustrate the importance of measuring the amplitudes of the unemployment rate. Amplitude measures, in general, by quantifying an important aspect of cyclical behavior, facilitate description, comparison, and understanding of past behavior, permit evaluation of current developments in terms of historical experience, and assist in the anticipation of future fluctuations in specific activities.

The amplitudes of expansions, contractions, and full cycles in economic time series can be measured in a variety of ways. For example, they might be measured in original units, that is, in terms of absolute differences between standings at peaks and troughs. They might be measured as percentage changes between two adjacent turning points, as the difference between adjacent turns relative to their average or relative to some other base. In the selection of amplitude measures, it is desirable that such measures be comparable among different series, among different cycles of the same series, and among different cycle phases. Furthermore, amplitudes should be measured in toto as well as per time unit, and the measures should be reasonably free from the effect of random movements.

The National Bureau's standard business cycle analysis meets most of these specifications by expressing the difference between peaks and troughs in relation to the cycle average, by measuring amplitudes both as total rise or fall and as amplitudes per month, and, wherever possible, by measuring amplitudes between three-month averages centered on peaks and troughs. Within this framework, there are still a variety of amplitude measures that can be computed. They may be relative or absolute, i.e., expressed in terms of ratios to the cycle base or in terms of deviations from the cycle base. They may be measured between standings at business cycle turns (reference cycle amplitudes) or between actual peaks and troughs of the analyzed activity (specific cycle amplitudes). Finally, they may be measured by reference to trough-to-trough or to peak-to-peak cycle bases. These various measures, all of which are part of the programmed business cycle analysis, are discussed below.

Computation of Amplitudes. Since the derivation of each of the several amplitude measures is similar in principle to every other, as is the format of the amplitude tables, it will suffice to analyze one output table in detail, showing the derivation of the various measures, interpreting the output, and discussing the problems that may arise. Let us use Output Table 3A-2, which shows relative reference cycle amplitudes for trough-to-trough cycles in employment. Relative amplitudes, that is, amplitudes relative to the cycle base, are the measures provided by the standard analysis.[14] That reference cycle analysis is being dealt with can be seen from the notation "reference cycle analysis" in the upper left-hand corner of Output Table 3A-2. The

[14] If the whole analysis is on an "absolute" basis (requiring a different program), this is indicated by the words "absolute changes" in the title of the first table of the analysis and by the words "cycle deviations" in the amplitude tables.

trough-peak-trough base for the cycle measures, finally, is indicated by the heading of the table, "Cyclical Amplitudes, Trough-to-Trough Analysis," and by the arrangement of the stub (extreme left-hand column), which gives in each line the three dates that delineate a TPT reference cycle. (The reference cycle chronology containing these dates must be provided as input for the electronic analysis.)

The first three columns of Output Table 3A-2 provide cycle relatives of the average values of the standings of the series at three reference turns. Ordinarily, a three-month average of the seasonally adjusted data (Output Table 3A-1) is computed and printed in the relevant trough-peak-trough columns I, V, and IX of Output Table 3A-3 (which is called "Cycle Patterns, Standings in Original Units").[15] These average standings at troughs and peaks are divided by their cycle base, that is, by the average of all original values included in the particular cycle. This cycle base is given in the last column of Output Table 3A-3, and the average standings expressed as cycle relatives are given in columns 1 to 3 of Output Table 3A-2. From these standings at reference cycle turns, amplitudes (columns 4, 5, and 6) are computed. The *rise* is computed by subtracting the standing at the initial trough from that at the peak, the *fall* by subtracting the peak standing from that of the terminal trough, and the *total* amplitude by adding the two movements, i.e., by subtracting the amplitude of the fall (including the sign) from that of the rise (column 4 minus column 5).[16] Under these definitions, total reference cycle amplitudes may be positive or negative, depending on the direction and magnitudes of changes during reference expansions and contractions. On the other hand, total specific cycle amplitudes (Output Tables 3A-22 and 3A-30) will always be positive for positively conforming and negative for inverted series.[17]

[15] If the series begins or ends at a reference turn, the program automatically computes a two-month average and prints it as an estimate of the missing month at the end of Output Table 3A-1. If it is desired to substitute two-month averages or the value of the middle month for three-month averages (as might be done in case of step functions or presmoothed series), such substitution must be specified by the exercise of certain options which will be described in a later section.

[16] The amplitudes are derived by simple subtraction because the standings are expressed as percentage relatives, with the cycle base equaling 100 per cent.

[17] Reference cycle amplitudes can be computed for series that do not show actual cyclical reversals in direction. These series may portray activities with strong secular trends that experience cyclical decelerations only. Even without

Columns 7, 8, and 9 of Output Table 3A-2 give amplitudes per month. These are derived by dividing the full amplitudes by the corresponding phase durations (as given in Output Table 3A-9). The amplitudes per month are rate-of-change measures, showing the vigor of the expansions and the sharpness of the contractions. These measures help to describe and to compare the impact of cyclical changes during short and long cycle phases. Both full amplitudes and amplitudes per month are analytically important. While the largest total fall in employment during the period under review occurred during the 1937–38 business contraction, by far the sharpest monthly fall happened during the demobilization period at the end of World War II.

The bottom lines of amplitude tables such as Output Table 3A-2 contain some summary measures that should be briefly explained. The total is important only computationally—as a step toward the average given in the next line. The print-out of the totals is convenient in case one wishes to modify the summary measures, say, by omitting war cycles or by adjusting the time period in order to facilitate comparison between different activities. The averages are, of course, of utmost importance, since they provide an approximation to the typical amplitude of the analyzed series. On the average, employment in nonagricultural establishments tended to rise 18.6 and to fall 5.5 per cent (of the cycle base) during business cycles in the period covered. In case of amplitudes per month, both simple and weighted averages are provided. For the latter the durations of the phases covered, in months, serve as weights. The average deviation, finally, provides a measure that permits gauging the representativeness of the average amplitude measures and the dispersion around them. It is computed by taking the mean of the absolute deviations, i.e., the differences between the amplitude of each phase and the mean of all phases.

The deviation measures call attention to the fact that scanning of summary measures is no substitute for detailed examination. The sum-

trend adjustment, differential behavior during business cycle expansions and contractions can be ascertained by comparing a series' reference phase amplitudes; and cyclical volatility can be measured by full-cycle amplitudes. Amplitudes for any given phase must always be compared with those of the preceding and succeeding phases; full-cycle amplitudes must be measured on both a TPT and a PTP basis. This is necessary so that the difference between contraction amplitudes and those of the preceding expansions is not regarded as cyclical, while it may, in fact, be due to a secular trend with a declining rate of growth.

mary measures do not reveal the tendency toward decreasing amplitudes of rises and falls in employment during the thirty years under review. Nor do they show to what extent the magnitudes, and perhaps the trend of the amplitudes, were influenced by the 1938–45 war cycle. It is the possibility of studying the individual measures, of adjusting the averages for nontypical events or for differences in time coverage, that makes the voluminous output of computer runs so valuable for research and analysis.

Relative and Absolute Amplitude Measures. The amplitude measures just discussed are part of an analysis in which most measures and all changes are expressed as relatives of their cycle base. This procedure is the most common version of business cycle analysis. However, there are situations in which this approach is not the only possible one; nor is it necessarily preferred or even feasible. An alternative consists of expressing cycle standings as absolute deviations from, rather than as relatives to, the cycle base and measuring amplitudes (and other changes) as differences between these absolute deviations in original units. This is equivalent to using stage standings in original units, without reference to the cycle base. The expression of standings as deviations from cycle averages, however, may be convenient for some analytical purposes.

Let us briefly review the conditions under which it might be preferable to compute absolute changes. One group of cases is that of rates, such as rates of unemployment, interest rates, or perhaps rates of change. In these cases the observations are already in relative terms, although the base for these relatives may be another variable. This permits intercycle comparisons of amplitudes in terms of absolute differences without converting the data to relatives of the cycle base. In other cases the conversion of rates into relatives of their cycle base may not be particularly meaningful, or both types of measures may be desired. For example, if components of GNP are being analyzed, it may be desirable to have analyses both in dollars and in relatives, since the effect of each component upon the aggregate can be measured in dollars. Another reason is operational: If the original data contain negative values, the computation of cycle relatives may not be feasible.[18]

[18] It might be feasible if the negative values disappear in the averaging process involved in the computation of stage standings. However, if there are several negative values close to each other, they may not average out, especially if they become the locations of specific cycle troughs.

A business cycle analysis of unemployment rates, in terms of absolute changes, is presented as Appendix 3B. It should be remembered that the unemployment rate is an inverted series; that is, its increases match business cycle contractions and its decreases match business cycle expansions. Again, Output Table 3B-2 contains reference cycle amplitude measures. Note, however, that the first three columns are cycle deviations (instead of cycle relatives),[19] and that hence negative values appear (in this case, because of the inverted behavior, in the peak column). The amplitudes are again derived by subtracting the initial trough from the peak standings for the rise column, and the peak from the terminal trough value for the fall column.[20]

Before substantive results of the amplitude computations are discussed, it should be remembered that the data are printed in units of per cent times 100. The amplitude measures show that between 1933 and 1937 the unemployment rate was reduced by about 14.8 percentage points (from 27.5 per cent to 12.7 per cent of the labor force), and that it rose between 1960 and 1961 by about 1.5 percentage points. The summary lines of the amplitude table demonstrate particularly forcefully the inadequacy of regarding average amplitudes (or other average measures) as typical, without examining dispersion and component behavior. In case of rises the average deviation (in either direction) of amplitudes from their average is 7 percentage points—almost as large as the average amplitude itself. And the component amplitudes range from a decrease of 20 per cent (relative to labor force totals) during 1938–45 to an increase of .02 per cent from 1945 to 1948. Examination of measures during individual cycles shows also the large differences between the amplitudes before and after World War II. This must of course be understood in terms of the general mildness of recent cycles.

The differences in cyclical measures resulting from the analysis of absolute changes and the analysis of relative changes can be illus-

[19] These deviations are derived from columns 1, 5, 9, and 10 of Output Table 3B-3, as are the cycle relatives of the other version.

[20] This requires strict regard to signs. Subtracting the negative peak deviation in 1937 (−615.2) from the positive trough deviation in 1938 (+291.5) results in a positive amplitude during the fall (+906.7). Note also that the rise column, which contains mostly negative values (i.e., decreases in the unemployment rate), also has one positive value (20.0) for the business cycle expansion of 1945–48. This means, of course, that the unemployment rate rose slightly between the immediate postwar trough and the business cycle peak of 1948.

trated by applying both analyses to the unemployment rate. The absolute amplitudes listed in Table 8 show, for instance, that during the recovery from the Great Depression unemployment declined by about 15 percentage points. This amounts to about three-fourths of the unemployment decline (20.6 points) during the next expansion in general business activity—from 1938 to the business cycle peak of World War II (see column 1). By contrast, the ratio of the two corresponding relative measures was only about one-third (78.5/229.1, column 4). This difference between the absolute and the relative measures is due to the change in the base of the relatives. During the first cycle, which includes the recovery from the Great Depression, unemployment was still comparatively high. Thus, the change from trough to peak, expressed relative to this cycle base, appears to be small. By contrast, the very low level of unemployment, during the war cycle,

TABLE 8

REFERENCE CYCLE AMPLITUDES IN THE UNEMPLOYMENT RATE, ABSOLUTE AND RELATIVE CHANGES, 1933–61

Reference Cycle Dates			*Absolute Changes (per cent of labor force)*			*Relative Changes (per cent of cycle averages)*		
T	*P*	*T*	*Rise* (*1*)	*Fall* (*2*)	*Total* (*3*)	*Fall* (*4*)	*Rise* (*5*)	*Total* (*6*)
3/33	5/37	6/38	−14.8	+9.1	−23.8	−78.5	+48.2	−126.6
6/38	2/45	10/45	−20.6	+2.6	−23.1	−229.1	+28.8	−257.9
10/45	11/48	10/49	+0.2	+3.0	−2.8	+4.6	+69.4	−64.8
10/49	7/53	8/54	−4.4	+3.3	−7.7	−108.2	+81.8	−190.0
8/54	7/57	4/58	−1.7	+2.9	−4.6	−36.9	+63.5	−100.5
4/58	5/60	2/61	−1.9	+1.5	−3.4	−31.7	+25.6	−57.3
Average			−7.2	+3.7	−10.9	−80.0	+52.9	−132.9
Average deviation			7.0	1.8	8.4	59.1	18.7	60.7

Note: Unemployment shows an inverted cyclical relationship, i.e., specific cycle peaks correspond to reference troughs and specific cycle troughs to reference peaks. Rise and fall denote expansions and contractions in general business conditions. Total equals rise minus fall.

caused the next cycle base to be low and, hence, changes relative to that base to be large.

Another illustration of the different results obtained from absolute and relative measures is provided by comparison of the 1957–58 and 1937–38 contractions. The absolute analysis shows the rise of the unemployment rate during 1957–58 to be only a third as great as that during 1937–38 (2.9 vs. 9.1). In relative terms the rise in 1957–58 is greater than that during 1937–38 (63.5 vs. 48.2). Note also that, in the case of absolute changes, the average deviation of the total amplitude is about 77 per cent of the average over all cycles (column 3); the corresponding figure for the relative analysis is 46 per cent (column 6). The importance of such differences can only be evaluated in the context of a particular research problem. The object of the preceding illustration is to point to the availability of the two analyses and to the differences in the measures generated.

Reference and Specific Amplitudes. All amplitude measures considered so far describe changes during expansions and contractions in general business activity; that is, they are reference cycle measures. These amplitudes are characteristically smaller than the full cyclical swings of an economic activity between its own peaks and troughs. The differences in amplitudes are caused by differences in the dates of specific turns and business cycle turns. When these turns coincide, as they do occasionally, the reference cycle and specific cycle amplitudes are, of course, identical. When they do not, the reference cycle amplitude describes less than the full change between the highs and the lows of the series itself.

The amplitude measures describing changes between the turning points established for specific activities are called specific cycle amplitudes. They are found in the section of the computer runs headed "specific cycle analysis." There are two amplitude tables in that section—Output Table 3B-22 for trough-to-trough and Output Table 3B-30 for peak-to-peak analysis. The derivation of these tables is strictly analogous to that described for reference cycles except, of course, for the selection of turns.

Specific cycle amplitudes cannot be readily computed if series do not show actual cyclical increases and declines. In such cases, some trend adjustment is necessary if specific cycle analysis is desired. This may be particularly important for data with cycles that cannot be readily related to an existing reference chronology (foreign data, pre-

1850 data, data with atypical changes in growth rates). There are several ways to determine cyclical characteristics of such data: computation of deviations from trend, derivation of measures of change and rates of change, or establishment of cyclical steps in first differences and growth rates. None of these procedures is part of the National Bureau's standard business cycle analysis. However, once specific cycle turns or their equivalents are established, the programmed specific cycle analysis can be applied.

Comparison between Output Tables 3B-22 and 3B-2 and Output Tables 3B-30 and 3B-10 shows, as expected, that for every comparable cycle phase specific amplitudes are larger than, or at least as large as, the corresponding reference cycle amplitudes. The difference is rather small in case of employment, since the timing difference between specific and reference cycles is small; in the unemployment rate it is somewhat larger, since here the turns deviate more markedly from those in general business conditions. It is again important to watch the comparability of the printed summary measures, which are reproduced in Tables 9 and 10. In the case of the employment series, there is no problem. Amplitude averages for reference and specific cycles refer to similar time periods, and, as expected, the summary measures show reference cycle amplitudes to be smaller than specific cycle amplitudes. For the unemployment rate, however, in case of peak-to-peak analysis, average specific cycle amplitudes are smaller than average reference cycle amplitudes—for both cycle phases and for the full cycle. The cause for this seeming anomaly lies in the different coverage of years and cycle phases. As can be seen in Table 10, reference cycle amplitudes, on a peak-to-peak basis, are averaged over the span 1929–60, while those for specific cycles are averaged over the years 1937–59. Thus neither the Great Depression nor the subsequent recovery is included in the specific cycle averages. Their exclusion readily explains the smaller average for specific cycles.

Monthly specific cycle amplitudes are not necessarily larger than monthly reference cycle amplitudes, even if corresponding cycle phases are compared. Examples of such occurrences can be found in monthly amplitudes of falls for the two illustrative series in the relevant output and summary tables. This situation is not particularly surprising. Per month amplitudes are ratios of phase amplitudes to phase durations. If the durations are long, per month amplitudes become correspondingly smaller. Phase durations are a consequence of the location of

TABLE 9

AVERAGE AMPLITUDES IN NONAGRICULTURAL EMPLOYMENT, RELATIVE CHANGES

Output Table No.		*Standings in Cycle Relatives at Business Cycle Turns*				*Amplitudes during Business Cycle*				*Amplitudes per Month during Business Cycle*			
		T	*P*	*T*	*P*	*Rise*	*Fall*	*Rise*	*Total*	*Rise*	*Fall*	*Rise*	*Total*
		REFERENCE CYCLES (NO RECOGNITION OF TIMING DIFFERENCES)											
3A-2	Trough-to-trough (1933–61)	+88.0	+106.5	+101.0		+18.5	−5.5		+24.1	+0.38	−0.54		+0.40
3A-10	Peak-to-peak (1929–60)		+100.4	+89.7	+108.7		−10.7	+19.0	−29.7		−0.58	+0.39	−0.44
		REFERENCE CYCLES (TIMING DIFFERENCES RECOGNIZED)[a]											
3A-18	Trough-to-trough (1933–61)	+88.0	+106.5	+101.0		+18.5	−5.5		+24.1	+0.39	−0.49		+0.40
3A-20	Peak-to-peak (1929–60)		+100.5	+89.8	+108.8		−10.7	+19.0	−29.7		−0.54	+0.40	−0.44
		SPECIFIC CYCLES											
3A-22	Trough-to-trough (1933–61)	+88.2	+107.1	+101.2		+18.9	−5.9		+24.8	+0.41	−0.42		+0.41
3A-30	Peak-to-peak (1929–60)		+101.5	+90.5	+110.0		−11.0	+19.5	−30.6		−0.49	+0.43	−0.45

[a] Trough not shifted, peak shifted by −1 month.

TABLE 10

AVERAGE AMPLITUDES IN THE UNEMPLOYMENT RATE, ABSOLUTE CHANGES

Output Table No.		Deviations from Cycle Base at Business Cycle Turns				Amplitudes during Business Cycle				Amplitudes per Month during Business Cycle			
		T	*P*	*T*	*P*	*Rise*	*Fall*	*Rise*	*Total*	*Rise*	*Fall*	*Rise*	*Total*
		REFERENCE CYCLES (NO RECOGNITION OF TIMING DIFFERENCES)											
3B-2	Trough-to-trough (1933–61)	+4.38	−2.81	+0.93		−7.18	+3.74		−10.92	−0.13	+0.34		−0.17
3B-10	Peak-to-peak (1929–60)		−3.46	+4.50	−2.69		+7.95	−7.18	+15.14		+0.42	−0.13	+0.20
		REFERENCE CYCLES (TIMING DIFFERENCES RECOGNIZED)[a]											
3B-18	Trough-to-trough (1933–61)	+4.44	−2.69	+0.95		−7.13	+3.64		−10.77	−0.15	+0.22		−0.17
3B-20	Peak-to-peak (1929–60)		−3.23	+4.55	−2.58		+7.77	−7.13	+14.90		+0.29	−0.15	+0.20
		SPECIFIC CYCLES											
3B-22	Trough-to-trough (1933–61)	+4.70	−2.90	+1.21		−7.59	+4.11		−11.70	−0.17	+0.28		−0.19
3B-30	Peak-to-peak (1937–59)		−0.98	+3.57	−2.42		+4.55	−5.99	+10.54		+0.32	−0.14	+0.18

Note: This is an inverted series; specific peaks correspond to reference troughs, and specific troughs to reference peaks.

[a] Trough shifted by 2 months, peak by −4 months.

turning points, which, as was pointed out, is highly sensitive to small variations in level, subject to fringe decisions in case of double turns, and so forth. It follows that in computing average monthly amplitudes, large full amplitudes can be easily accompanied and compensated or overcompensated by long durations. It also follows that average monthly amplitudes are much more dependent on turning-point determination, and hence less stable, than the full-phase amplitudes.

Output Tables 3B-18 and 3B-20 and Tables 9 and 10 contain information on adjusted reference cycle amplitudes, i.e., on amplitudes after recognition of timing differences. These amplitudes are conceptually somewhere between simple reference cycle amplitudes and specific cycle amplitudes, in that they shift the reference dates at peaks and at troughs by the median lead or lag of the series at these turns. If the average timing showed no dispersion at all, these amplitudes would be identical with specific cycle amplitudes. This, of course, cannot often be observed. However, when leads or lags are only moderately dispersed, the adjusted reference cycle amplitudes are generally larger than ordinary reference but smaller than specific cycle amplitudes. In so far as the leads or lags are systematic, the adjusted amplitude represents the extent of the series' reaction to business cycles better than the other measures of amplitude do.

Table 9 shows that for nonagricultural employees, the amplitudes before and after recognition of timing differences are the same. This is because the series has virtually coincident timing. Table 10 shows distinct differences between the amplitudes of unemployment rates before and after recognition of timing differences (Output Tables 3B-2 and 3B-18, 3B-10 and 3B-20). However, the differences are not in the expected direction—that is, the average amplitudes after recognition of timing differences are smaller than before such recognition, although the same time period is covered. To understand why this may happen, let us compare the individual entries in the rise columns of Output Tables 3B-2 and 3B-18. The first table requires no comment. The latter table contains the notation "timing diffs. recognized by shifting ref. dates, trough 2 months, peak −4 months." These are the median number of months by which, over the period of observation, cyclical turns in the unemployment rate tended to precede peaks and to lag behind troughs in general business activity. The relatively smaller amplitudes, after recognition of timing differences, are brought about by the dispersion of individual leads and lags around their median. At the 1949 trough, for example, the peak in unemployment (7.8 per cent) coincided with the business cycle trough, while two months later (the interval of the median lag) unemployment had dropped considerably (to 6.8 per cent). Since the alternative trough values of unemployment (at the business cycle peak and four months before the peak) were about the same, the change during the actual reference phase was larger than that during the adjusted reference phase. This, together with a similar occurrence around the 1937 business cycle peak, caused the average adjusted amplitudes to be smaller than average reference cycle amplitudes. Such unexpected decreases in amplitude after adjustment for average timing differences are most likely to occur if the average lead or lag is relatively short and the dispersion is large.

Some peculiar problems can arise in the computation of amplitudes after shifting reference turns for typical timing behavior. Since the shifting is done by average (median) monthly leads or lags at peaks and at troughs, the shift in a given cycle may lead to an inversion of the proper order of turns; that is, if a trough is shifted forward and the subsequent peak backward, the shifts can overlap so that the peak is shifted to an earlier date than the trough, and the cycle phase disappears. In such a case the shifting does not make sense, nor would the computed amplitudes. If such overlapping occurs, the program omits computation of average amplitudes and indicates this fact, in footnotes to Output Tables 18 and 20. It is not clear, however, that actual overlapping is the proper criterion for disregarding amplitudes. The results may become valueless also if, after adjustment for average timing, the reference phases shrink to one, two, three, or so months. Whether such short phases make sense depends on the duration of the reference cycle, and perhaps also on that of the specific cycle. Thus, the program provides Output Table 3B-19, which gives durations for each adjusted reference cycle phase. This table, then, makes it possible to evaluate the meaning of the amplitudes, after recognition of timing differences, presented in Output Table 3B-18. The problem of overlapping phases could be avoided if the median lead or lag, and hence the shifting of reference dates, were expressed as a percentage of the average cycle phase rather than in months. This procedure would correspond more closely to the division of cycles into stages in computing reference cycle patterns.

The possibility of using alternative cycle bases (TPT cycles or PTP cycles) was discussed in the general description of the National Bureau's approach to business cycle analysis. Most previous illustrations were based on TPT output tables and termed "Trough-to-Trough Analysis," which corresponds to what is called "positive plan" in Burn's and Mitchell's *Measuring Business Cycles* and in some earlier versions of business cycle analysis. The alternative amplitude computations are contained in the section called "Peak-to-Peak Analysis," which corresponds to what Burns and Mitchell called "inverted plan." Reference cycle amplitudes on a PTP basis are given in Output Tables 3A-10 and 3B-10 (straight chronology) and 3A-26 and 3B-26 (timing differences recognized); specific cycle amplitudes are found in Output Tables 3A-30 and 3B-30. The results can be conveniently compared with those derived from TPT analysis, on the basis of the summary measures offered in Tables 9 and 10. For present purposes it suffices to state that the amplitude measures derived by use of alternative cycle bases are rather similar. This similarity provides some insurance that the type of cycle base chosen did not exercise any undue influence on research results. Also, when amplitudes of successive cycle phases are being compared, it may be desirable to have the amplitudes computed on the same base, in which case both sets of measures are needed.

Perhaps the statement bears repetition that the PTP and the TPT analysis do *not* cover the same period. Only full TPT or PTP cycles are included in the computation of amplitude and other measures (see Chart 10). Thus, there may be odd phases left at the beginning or end of periods that could be included in the phase averages if desired.

CYCLE PATTERNS

Use and Concept. Up to this point, all the analytic measures discussed have pertained to turning points—their identification and dating, their timing relation to business cycle turns, the levels which a given time series assumes at specific and reference turns, and the changes between these levels. This concentration of attention on the few, admittedly important, moments at which time series change their cyclical direction disregards all information about cyclical behavior between turning points. The analytical measures used to summarize such intraphase behavior are the so-called cycle patterns. As briefly described before, each cycle is divided into nine stages. Three-month averages around initial trough, peak, and terminal trough form stages I, V, and IX, and the intermediate stages are averages for thirds of expansions (II, III, IV) and thirds of contractions (VI, VII, VIII).[21]

The uses of cycle patterns are manifold. The standardized summarization of cyclical behavior serves descriptive and analytical purposes, such as comparisons between cyclical characteristics of different industrial activities or between different cycles of the same activities. The question of whether cycle patterns of the same activity show stable characteristics can be studied on the basis of such measures. Cycle patterns may also prove useful for the testing of broad theoretical hypotheses on cyclical behavior. Are the cycles of major industrial activities sinusoidal, and do the nine-stage patterns reflect sinusoidal characteristics such as points of inflection? Do investment series show the rapid (and synchronous) declines that may be expected from hypotheses such as that of the collapse of the marginal efficiency of capital? Is it true that the periods of early recovery and late boom are characterized by particularly rapid changes and that consequently the patterns of expansion show a leveling off in the middle—a phenomenon sometimes labeled "mid-expansion retardation"? There may also be more practical applications of cycle patterns. If the patterns of a particular activity show, in most cycles, sufficiently homogeneous features—such as early spurts or slow starts—these regularities can be used in the interpretation of current behavior and as a guide to impending events.

While the measures discussed so far (timing, duration, amplitudes) are rather straightforward and the assertion of their relevance is not based on any particularly controversial assumptions, this is less true for the cycle patterns. The basic problem is, of course, whether fixed fractions of cycle phases of varying chronological lengths (such as thirds of expansions) are analytically meaningful units. (This is apart from the question of whether the tripartite division is the best possible one.) Consider the recent expansion starting in 1961. If it had ended around mid-1962, a supposition which is not beyond reason, stage III would have lasted about six months and extended from about August 1961 to about January 1962. If it ended in November 1969, stage III would have lasted thirty-four months and extended from January 1967

[21] Technical detail, such as the distribution of cycle phases that are not divisible by three, is handled by the program. The rule for this particular decision is that phases III and VII are lengthened or shortened to absorb odd months.

to October 1969. If one concedes that the end of an expansion is not fully predetermined but is influenced by many economic or noneconomic factors that develop during the expansion itself, it is difficult to assume that stages, the delineation of which depends on the eventual location of cyclical turns, are homogeneous. On the other hand, the periods immediately before and after cyclical turns must be expected to have characteristics different from, say, stages III or VII, no matter how long the phase, and one way to find these characteristics is to segment phases in accordance with the National Bureau analysis. Stage measures of some kind are also necessary for a systematic investigation of how the cyclical characteristics of phase segments are most efficiently described.[22]

Reference and Specific Patterns. The business cycle program yields a variety of measures pertaining to cycle patterns. Information on intraphase behavior is provided in the output tables of Appendixes 3A and 3B. The table numbers are listed below.

Measures	*Reference Cycles*		*Specific Cycles*	
	TPT	PTP	TPT	PTP
1. Standings in original units	3	—	23	—
2. Standings as cycle relatives or as deviations from cycle base	4	11	24	31
3. Stage-to-stage change of standings, total change	5	12	25	32
4. Stage-to-stage change of standings, change per month	6	13	26	33
5. Intervals between midpoints of cycle stages	7	14	27	34

Measures 2, 3, and 4 are, analytically, the most important; 1 and 5 are mainly required in the derivation of the former. To demonstrate the type of knowledge that can be extracted from these patterns, emphasis will be placed on the trough-to-trough patterns for 1933–38 and 1958–61. Chart 11 shows these patterns for the series on employ-

[22] The measurement of cyclical changes during chronological stretches of given length (in months) will be discussed in Chapter 4. Segmentation could, in principle, also be based on other criteria such as inflection points, points of maximum diffusion, attainment of previously experienced levels, relations to other series, and so forth.

CHART 11

REFERENCE AND SPECIFIC CYCLE PATTERNS DURING TWO CYCLES, NONAGRICULTURAL EMPLOYMENT AND UNEMPLOYMENT RATE, 1933–38 AND 1958–61

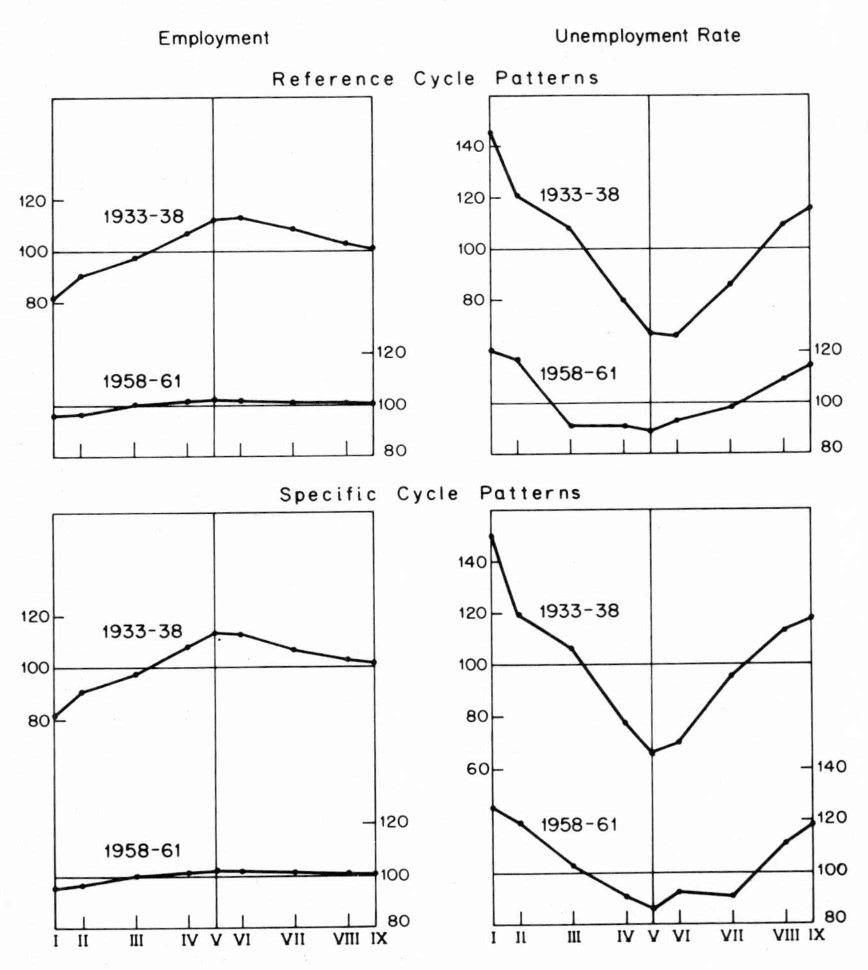

ment and on the unemployment rate. To make the measures comparable, cycle relatives are used for both series.

Horizontal distances between stage standings are standardized, without reference to chronological stage duration.[23] The upper panel of the chart contains reference cycle and the lower panel specific cycle patterns. Some of its features, such as the inverted character of the unemployment rate, the larger relative amplitudes of unemployment compared with employment, and the milder fluctuation during the postwar as compared to the interwar cycle, are also apparent from the amplitude tables. Note the smoother contour of employment as compared with unemployment cycles and the greater resemblance between reference and specific patterns in case of nonagricultural employment (which of course is due to the practically coincident timing). Note also that the peculiar timing characteristics of the unemployment rate (its lag at troughs and lead at peaks) become obscured by the process of averaging over cycle stages, and so do the traces of midexpansion retardation that may be found in the original data (see Chart 1 in Chapter 2). Actually, the cycle patterns contained in Chart 11 do not show any dramatic deviations from a simple triangular pattern shared by both cycles. However, two cycles are certainly not enough to detect pervasive characteristics, and the activities chosen, because of their nature and broad coverage, are not likely to display strong intraphase idiosyncrasies. Illustrations of such idiosyncrasies can be found in most collections of cycle patterns.[24]

A few comments are in order on stage-to-stage changes as found in the eight output tables listed opposite items 3 and 4 above. Being first differences, these changes tend to display considerably more instability than the cycle relatives from which they were derived. For the same reason, they emphasize the comparative lack of smoothness in the unemployment patterns. Nevertheless they show, particularly in the form of monthly changes, a tendency of the unemployment rate to come down sharply at the beginning of business expansions but to change more slowly after peaks. This difference between early expan-

[23] This is not the only way to chart cycle patterns. Burns and Mitchell, *Measuring Business Cycles,* contains many illustrations in which cycle patterns are charted on scales that represent the chronological duration of phases and stages (see, for example, pp. 155 and 165).

[24] See, for example, Burns and Mitchell, p. 173; Mitchell, *What Happens During Business Cycles,* pp. 32–49; Moses Abramovitz, *Inventories and Business Cycles,* New York, NBER, 1950, pp. 274, 282–283, *et passim.*

sions and contractions is less pronounced in the stage-to-stage changes of employment patterns.

Average Pattern. The division of the cycle into stages makes it possible to average stage standings over cycles and to compute average cycle patterns that may be expected to describe some pervasive cyclical characteristics of a given variable.

These average patterns must be used with great care. They are averages (over several cycles) of averages (stage standings) that refer to periods of different duration and cycles of widely varying amplitudes; they are averages composed of relatively few cycles, so that a single large fluctuation (such as the 1929–37 cycle) may dominate the average pattern; and they are averages that may cover a varying number of phases and cycles. From all this, it follows that average patterns must always be scrutinized for representativeness, coverage, and comparability, and that they should be used only in conjunction with component cycles. On the other hand, average patterns do bring out common characteristics and can be valuable for comparative analysis. R. A. Gordon, for instance, used cycle patterns very effectively to compare cyclical behavior of economic activities before and after World War II.[25] He found important differences in cyclical characteristics. Note also that the average cycle patterns based on the postwar experience are clearly more representative of their component cycles than the prewar average patterns are of theirs.

CONFORMITY MEASURES

General. In the cyclical analysis of economic activities, an obvious concern is the way in which fluctuations in any particular variable are related to fluctuations in business conditions at large. Are the fluctuations of the variable positively or invertedly related to business cycles? Can every specific cycle be matched with a reference cycle? Are there extra cycles or have some been skipped in the particular activity? How close is the correspondence of cycles in a specific activity and in general business conditions? These questions are not only of interest for the analysis of particular variables, but their resolution for many variables indicates the degree of consensus of cyclical movements in different sectors of the economy and in the economy at large.

Some information about the correspondence of specific cycles to business cycles—that is, information about their conformity—can be

[25] *Business Fluctuations,* New York, 1961, pp. 265 ff.

extracted from several of the approaches and measures already discussed, such as inspection of time series plotted against a reference cycle grid, comparisons of timing and duration measures for specific and reference cycles, or comparisons of specific and reference cycle patterns. However, explicit measures describing conformity are desirable. Indeed, conformity measures become indispensable if the behavior of whole groups of activities is to be summarized. The question of the relative conformity of agricultural as compared to industrial prices can hardly be answered without some systematic approach to the measurement of conformity, such as the approach of the National Bureau's analysis.

Conformity measures are designed to describe and summarize the differential behavior of given economic activities during business cycle expansions, contractions, and full cycles and to establish whether the cyclical fluctuations of given activities are characteristically related positively, invertedly, or not at all to business cycles. They are also designed to measure the degree of conformity and to measure changes in this degree when typical leads and lags in the cyclical turning points of given activities are taken into account.[26]

Conformity Without Regard to Timing Differences. Conformity measures are part of the reference cycle analysis; in the specific cycle analysis, time series are not related to business cycles at large. The meaning of these measures will be explained on the basis of Output Table 3A-17 of the programmed analysis of nonagricultural employment (p. 143). The upper part of the table contains average changes per month, in terms of reference cycle relatives, during expansions and during contractions—in the left half of the table for trough-to-trough cycles and in the right half for peak-to-peak cycles. Note that during every business cycle expansion there was an increase in employment and during every contraction there was a decline, indicating a clear-cut case of perfect positive cyclical conformity. This is expressed by the index of 100 for all expansion, contraction, and full-cycle conformities. Conformity indexes are derived in Output Table 3A-17 as described below.

1. Indexes of *expansion conformity* are computed by counting the number of increases during reference expansions (procyclical movements), subtracting the number of decreases (countercyclical move-

[26] For a more detailed description and interpretation of these measures, see Burns and Mitchell, pp. 31–38, 176–197.

ments), and expressing the difference as a percentage of all entries including no changes. The result is the expansion conformity index, shown (in the row marked "expansions") for TPT cycles under the expansion column on the left, for all covered expansions in the middle column, and for PTP cycles under the corresponding column on the right of the table. In the present case, both expansion indexes indicate 100 per cent conformity. The use of net indexes (deduction of decreases from increases) permits the construction of a measure that fluctuates around zero, with negative figures indicating inverse conformity.[27]

2. Indexes of *contraction conformity* are computed analogously. Here the number of increases during business cycle contractions (countercyclical movements) are deducted from the number of decreases (procyclical movements), resulting in an index that is positive if the series generally falls during business cycle contractions.

3. *Full-cycle conformity* measures are based on the difference between the average monthly changes during expansions and during preceding or subsequent contractions. The difference is computed as contraction change minus expansion change (signs considered), which implies that positive conformity is indicated by a negative sign and vice versa. The difference is reported for TPT cycles in the column headed "con. minus preced. exp.," and for PTP cycles in the column headed "con. minus succed. exp." If contraction changes are negative and expansion changes positive, as in the present case, the resulting difference will be a negative entry larger than the average change in contractions or expansions alone. In series with strong trends, conformity with business cycles may be reflected in differential rates of growth (milder monthly increase in contractions in case of an upward trend) rather than in actual increases and decreases. If the response to business cycle contractions is just a deceleration of growth, the entries for contractions minus expansions would still be negative but smaller than the entries in either contraction or expansion alone. Full-cycle conformity indexes are computed by summarizing the consistency in the changes reported in the "con. minus preced. exp." and "con. minus succed. exp." columns. Again, the number of plus signs is subtracted

[27] The alternative—expressing the number of increases as a percentage of all entries—would lead to a measure fluctuating between 0 and 100. The percentage deviation of that measure from 50 would lead to the measure described in the text.

from the number of minus signs and the difference expressed as a percentage of all entries.

4. *Total full-cycle conformity* finally is measured as the weighted average (weighted by number of cycles covered) of the PTP and the TPT full-cycle conformity indexes. The total index is reported in the last line of the table. It can vary between +100 (total perfect positive conformity) and −100 (total perfect inverse conformity). An index of zero denotes absence of consistent conformity of a series to business cycles.

Nonagricultural employment presents such a straightforward case of uniformly perfect conformity that the corresponding table (Output Table 3A-17) of the reference cycle analysis of the unemployment rate should be examined also. Here most average monthly changes during expansions are negative and all changes during contraction positive, indicating inverse conformity. Consequently, the entries in the full-cycle column "con. minus preced. exp." are all positive (since unemployment rises more in contractions than in expansions) and, with the one exception of the 1945–49 cycle, numerically larger than those in the preceding two columns. This situation is reflected in the negative signs for the expansion, contraction, and full-cycle conformity indexes. One example of less than perfect conformity may be used for illustrating the derivation of the index. During the 1945–48 expansion, the unemployment rate went up a little, so that unemployment decreased in only five of six expansions. Hence, the conformity indexes for TPT cycle expansions, computed as described, is

$$\left(\frac{1-5}{6}\right) 100 = -67.$$

Note that the less than perfect (inverse) expansion conformity does not necessarily affect the full-cycle conformity. During the exceptional 1945–49 TPT cycle, the rise of the unemployment rate in the contraction was still stronger than the rise in the preceding expansion (left side of Output Table 3A-17); and during the 1945–48 TPT cycle, the rise of the rate during the contraction was stronger than that in the following expansion. Hence, in both cases the full-cycle entries were uniformly positive, leading to perfect inverse conformity indexes of −100. For a fuller understanding, one may consider what the conformity index would be if the entry during the 1945–48 expansion were 54.00 instead of .54. Then the monthly change for the full 1945–

49 TPT cycle would be −26.42 (i.e., 27.58 − 54.00) and for the 1945–48 PTP cycle −21.75 (i.e., 32.25 − 54.00). Consequently, the full-cycle conformity would be reduced to −66.7, reflecting the hypothetical condition that the unemployment rate not only rose in one of the six expansions but rose more than during the preceding and subsequent contractions.

Conformity Measures Recognizing Timing Differences. Suppose that the turning points of an economic time series led those of business conditions as a whole by about half a phase, or a quarter of a cycle. Then the conformity measures so far discussed may show little conformity despite the fact that each specific cycle can be readily matched with a corresponding reference cycle. This is so because the leading series may have declined so much by the time the reference peak is reached, and risen so much by the time the reference trough is reached, that there would be little difference between the two levels. This possibility points to the need for conformity measures that make allowance for the typical timing relation of the measured economic activity to business cycles. Burns and Mitchell developed an approach that involves determination of the business cycle stages during which an activity tends to expand and those during which it tends to contract. Graphs of business cycle patterns with standardized time scales [28] help to determine the reference stages of typical rise. Once these stages have been determined, conformity measures can be derived which are conceptually analogous to those previously discussed but are based on the changes of reference cycle relatives during phases of typical expansion and contraction. The typical expansion period is indicated in the table titles by noting that, for the analyzed series, expansions cover, say, stages II–VI and that, in case of positive conformity, specific expansions are matched with reference expansion.

For purposes of electronic computations, this procedure presents some drawbacks, since either the computations would have to be interrupted for the determination of the reference phases during which the series typically expands, or this determination would have to be formalized, possibly modified, and programmed. This would be a formidable task. Burns and Mitchell suggested alternative approaches,[29] among them one that took care of timing differences in terms of months rather than of cycle stages or cycle fractions. This alternative

[28] See Chart 8 and Burns and Mitchell, p. 187.
[29] *Ibid.*, p. 194.

was not pursued at the time, partly because the use of stages fitted the general analytical framework closely, and partly because the alternative involved recomputation of cycle averages, peak and trough standings, amplitudes, and so forth. The required computations, which are cumbersome if performed on desk calculators, are easily carried out with the aid of electronic computers. Thus, in the programmed version of the analysis, this general approach is employed.

Since nonagricultural employment shows perfect conformity in all measures and since the series is known to have virtually synchronous timing, the conformity measures, after consideration of timing relationships, can hardly be expected to differ much from those computed without consideration of timing. The relevant conformity table (Output Table 3A-21) shows, as expected, perfect positive conformity after consideration of typical timing—an average lead of two months at peaks and coincidence at troughs. The description of the programmed procedure will therefore be based on the analysis of the unemployment rate of Output Tables 3B-18 through 3B-21.

These tables, which are all part of the derivation of conformity measures after consideration of average leads and lags, carry the legend "timing diffs. recognized by shifting ref. dates, trough 2 months, peak −4 months." Note that the reference dates, rather than the time series itself, are shifted. This is done not only for computational convenience but of necessity because peaks and troughs are shifted by a different number of months, in accordance with the median timing of the series at the two types of reference turns. That is, since the unemployment rate tends to lag by two months at troughs and to lead by four months at peaks, reference troughs are uniformly shifted two months backward and reference peaks four months forward toward the specific turn, in order to compensate for the differential timing. The programmed reference cycle analysis is then applied, using the shifted reference dates.

New cyclical amplitudes, durations, and amplitudes per month are computed for TPT cycles (Output Tables 3B-18 and 3B-19), corresponding amplitude measures are derived for PTP (Output Table 3B-20), and new conformity measures are contained in Output Table 3B-21. The compensation for timing differences changes the unusual entry for the 1945–48 expansion from a small positive change (+.54 per cent of the civilian labor force) before recognition of timing differences to a small negative change (−1.56), indicating decrease of unemployment during this phase, after adjustment for typical timing. Consequently, all expansion entries have the same sign and the expansion conformity becomes perfect, on an inverted basis, for both TPT and PTP cycles.

The improvement of the conformity measures, after adjustment for timing differences, is plausible enough. However, the procedure is not quite as free from problems as may appear from our exposition. As noted earlier, if reference peaks are shifted forward and reference troughs backward, it can happen that the durations of the adjusted reference contractions become unreasonably short, become zero, or even become negative. The programmed procedure does not prevent such occurrence;[30] however, it offers a safeguard against unreasonable interpretation. By referring to Output Table 3B-19, which reports duration after adjustment for timing, the analyst can check for uncommonly short, zero, or negative durations. If these appear, these phases must be omitted, or the conformity indexes, after adjustment for timing differences, must be disregarded.

After adjustment for timing differences, conformity measures can also be lower than before consideration of such differences. All reference peaks are uniformly shifted by the median lead or lag of the given series, and the same is done for all reference troughs. This adjustment does not, of course, necessarily shift reference turns to specific turns. Since economic time series often have strong intraphase fluctuations, the uniform shift of, say, all peaks may lead in some instances to standings lower than those at the original reference peak. If the differences in levels are large enough to cause changes in the direction of movement, they may adversely affect phase conformity and total conformity. If these differences do not cause changes in direction but affect amplitudes relative to those of adjoining phases, they may still lead to changes in total conformity. Furthermore, the shift in timing affects the durations and thus the average monthly changes on which the conformity indexes are based. Thus, amplitude and duration changes may combine to bring about changes in conformity measures that may deviate from those expected on general grounds and possibly lead to lower conformity, after adjustment.[31] Such a deterioration of conform-

[30] If phases with zero and negative durations occur, no averages are computed.

[31] The effect of irregular intraphase movements on conformity can for analogous reasons also be positive, i.e., lead to spurious improvements in conformity indexes.

ity measures can usually be traced from the detailed output tables to particular phases and circumstances.

Concluding Remarks on Conformity. A conformity index can be regarded as a type of nonparametric correlation measure that describes the degree to which cycles in specific activities are associated with cycles in general activity. Similar to conventional correlation coefficients (except for the expression as percentage), the index varies between plus and minus 100, indicating a continuum from perfect positive to perfect inverse association. Like the correlation coefficient, the conformity index

> states the degree to which errors can be reduced in estimating the direction of movement of a series by taking account of its conformity to business cycles instead of guessing. . . . The greater the number of observations and the firmer the rational analysis, the greater our confidence in the significance of a coefficient of correlation, and so it is also with indexes of conformity. An index of conformity as low as +33 indicates that instances of positive conformity preponderate over instances of inverted conformity in the ratio of 2 to 1; but the index commands more serious attention when the cycles number ten than when they number three, and when they number thirty than when they number ten.[32]

The analogy suggests that conformity measures, like correlation coefficients, may be subjected to significance tests. Such tests have, indeed, been used in the evaluation of business cycle indicators.[33]

Conformity indexes as measures of correlation between cycles can also be employed when interest centers on the association between individual activities or groups of activities. The cyclical responsiveness (as distinguished from other shorter- or longer-term responses) of strike activity to unemployment rates, or of building activity to mortgage interest rates, may well be described by using the cyclical turns of unemployment or interest rates as a reference system and measuring the cyclical conformity of the related variables to this system. This is not to suggest that conformity indexes are the only or the best way to measure the cyclical aspect of the association. Nevertheless, they do focus on that aspect, and can be used to describe the behavior of one activity in relation to the specific cycles of another.

[32] Burns and Mitchell, p. 183.

[33] See Geoffrey H. Moore, *Statistical Indicators of Cyclical Revivals and Recessions,* New York, NBER, 1950; reprinted in Moore (ed.), *Business Cycle Indicators,* Vol. I, p. 206.

MEASURES OF SECULAR GROWTH

The standard business cycle analysis of time series does not provide measures of cyclical movements separate from longer-run trends. The analysis contains, however, measures that indicate changes from cycle to cycle, and thus intercycle trends. They describe changes between the cycle averages that form the bases for many of the cyclical measures in the National Bureau's analysis. These cycle averages, it will be remembered, are computed in several forms, for reference cycles as well as for specific cycles, for trough-to-trough as well as for peak-to-peak cycles; intercycle changes are computed for all these variants.

The output tables of Appendixes 3A and 3B that are relevant for the analysis of trends in employment and unemployment are:

Output Table	*Type of Cycle*
8	Reference, TPT
15	Reference, PTP
28	Specific, TPT
35	Specific, PTP

All these tables have average monthly standings, in original units, for each expansion, each contraction, and each full cycle. They also contain percentage changes between average standings for contiguous phases and contiguous cycles. The percentage changes between cycle averages are computed on two bases—as percentages of the earlier cycle and of the average of the two compared cycles. The latter measure has the advantage of avoiding the "percentage-base bias." The percentage changes between full cycles, finally, are computed as total and as average monthly changes. The tables also provide grand averages for the percentage changes (unweighted and weighted by duration) between adjoining full cycles.[34]

The secular measures can be utilized in a variety of ways. Average intercycle growth before World War I, during the interwar period, and during the period after World War II may be compared. The slackening of intercycle growth during recent postwar cycles in series such as

[34] For analyses based on absolute deviations, absolute changes rather than percentage changes are computed. This eliminates, of course, the need for two types of measures of cycle-to-cycle changes. See the output tables for analysis of unemployment rates.

nonagricultural employment can be observed—a finding, incidentally, that would have to be modified if the expansion that began in 1961 were included in the cyclical measures.

One can also combine the average phase standings or the average cycle standings into a time series of secular levels that can be charted and further analyzed. Some students have used overlapping TPT and PTP cycles, centered at the middle of the respective time spans.[35] Computation of percentage changes between overlapping cycles, incidentally, is an option available in the programmed analysis. For present purposes, it is not necessary to detail the numerous potential applications of the described secular measures to analysis of economic growth. The main point is that the various measures of average cyclical standings provide the student with levels that are reasonably free from short-cycle, seasonal, and irregular elements and thus permit him to measure and comment on long-term changes in economic activities in more detail and, perhaps, with more realism than that provided by the boldly general parameters of mathematically fitted trend equations.

MEASURES FOR QUARTERLY AND ANNUAL SERIES

Quarterly Series. Frequently time series of economic activities are available only in quarterly form. The national income accounts, many series pertaining to profits or the flow of funds, and many derived from anticipations and other surveys are examples. The standard business cycle analysis developed by Burns and Mitchell contains a variant especially designed for quarterly series.[36] In the computer program the need for a separate quarterly routine is avoided by converting all quarterly series to monthly form. Each quarterly figure is repeated three times—once for each month of the quarter. All turning points must be specified as occurring in the middle month of the quarter, and the chronology used for reference analysis must be a quarterly chronology (with quarters identified by the middle month). In other respects, the monthly analysis is performed as previously described. A few features that distinguish the programmed quarterly analysis from that of Burns and Mitchell are pointed out below.

The output measures are close to, but not exactly the same as those obtained by the quarterly Burns and Mitchell approach. The main

[35] Moses Abramovitz, *Evidences of Long Swings in Aggregate Construction Since the Civil War,* New York, NBER, 1964.

[36] Burns and Mitchell, pp. 197–202.

reason for the difference is that the patterns are derived on the basis of somewhat different interpolation rules.[37]

The amplitude measures obtained by the quarterly analysis, as programmed, are the same as those obtained by the Burns and Mitchell analysis of quarterly data. They are usually, but not always, smaller than the amplitudes obtained by the analysis of comparable monthly data (i.e., where the quarterly series is derived by summing the monthly series). This is because in the monthly analysis the turning point stages (I, V, and IX) are derived from three-month averages centered at the turns. If these turns are in the middle of the quarter, the resulting stage levels are the same as those derived from the programmed version of the quarterly analysis. If the turns are not in the middle of the quarter, the difference in levels is small. A simple illustration will clarify this point (panels A and B of Chart 12) and also the circumstances under which the amplitudes may deviate. Panel C shows why the conversion from the quarterly to the monthly series is done by a step function rather than by straight-line interpolation. Linear interpolation between the quarterly observations, coupled with three-month averaging at turning points, would obviously lead to smaller amplitudes.

It is highly unlikely that the individual leads and lags computed from quarterly data would be the same as those derived from comparable monthly data. This is so for two reasons: The timing of quarterly series is forced into multiples of three months, and the quarterly turn may be located in another quarter than the monthly turn (see Chart 12, panel D). However, there is no reason to assume that shifts of timing induced by the use of quarterly data would occur predominantly in one direction. Thus, the average timing of quarterly data can be expected to be fairly close to that of the corresponding monthly data, particularly if the average covers a long period. Experience shows that this is indeed the case.[38]

The user of the program for quarterly series should note the following technical points: (1) input cards in the required form (seasonally adjusted data, thrice repeated per quarter) can be obtained as output of the National Bureau's quarterly seasonal adjustment program; (2)

[37] For the interpolation rules used in the quarterly Burns and Mitchell approach, see *ibid.*, p. 199. In the programmed quarterly analysis the rules used are those of the monthly analysis.

[38] For illustrations, see *ibid.*, pp. 226–228.

CHART 12
AMPLITUDES IN PROGRAMMED
MONTHLY AND QUARTERLY ANALYSES

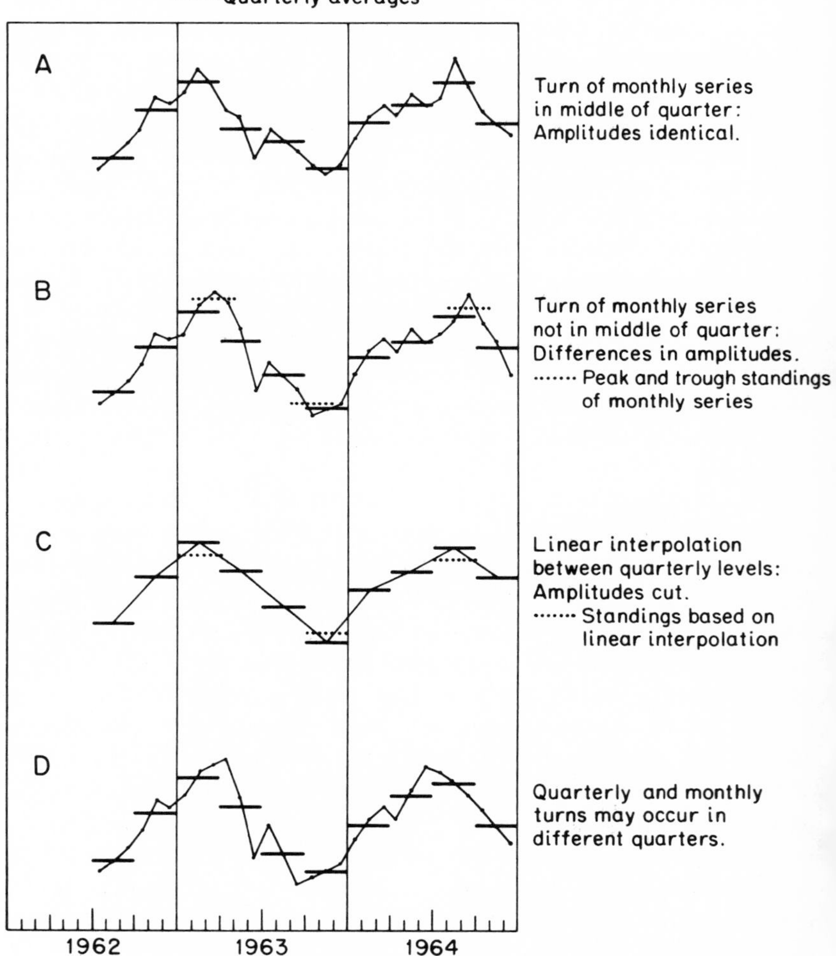

a quarterly, rather than a monthly, chronology must be used as reference cycle framework; (3) the median leads and lags for the B-4 analysis (conformity measures after allowance for timing differences) must be stated in multiples of three months to be compatible with the described procedures. In other respects, the technical instructions for monthly series can be applied to the business cycle analysis of quarterly time series after their conversion to monthly form.

Annual Series. As one goes back to economic statistics available before the turn of the century, monthly and quarterly information becomes rare and economic research often has to be based on information in annual form. Even today certain data (such as those presented in governmental budgets, the *Annual Survey of Manufactures,* some demographic surveys, and so forth) are available only in annual form. The question arises to what extent these annual series can be subjected to cyclical analysis.

It is clear that annual data present great handicaps for the derivation of useful cyclical measures. Some cycles may entirely disappear; leads and lags of others may be obscured, durations distorted, amplitudes averaged out, and cyclical patterns oversimplified. Nevertheless, as reference to Chart 13 shows, annual data can show cycles which may lend themselves to analysis.[39]

Most of the problems mentioned in the preceding paragraph are caused by the short duration of business cycles relative to the time unit of measurement. The problems are less severe if the major objective of the analysis is merely to measure apparent amplitudes and secular trends, or if the fluctuations to be analyzed are not the Mitchellian business cycles but longer fluctuations, such as the fifteen- to twenty-year swings in construction or long swings in rates of economic growth (Kuznets cycles).

The program treats annual data somewhat differently from monthly data. For instance, the annual observations themselves, rather than three-point averages around the turn, are used as standings at peaks and troughs. Averaging is unnecessary since, in annual data, short-term irregularities are already smoothed out. Another difference is that the program permits substitution of a five-stage pattern for the usual nine-stage pattern. The reason is, of course, that it may be difficult to partition short cycles containing a few annual observations into mean-

[39] *Ibid.,* Chapter 7.

CHART 13

CYCLES IN MONTHLY, QUARTERLY, AND ANNUAL SERIES, PIG IRON PRODUCTION, 1896–1933

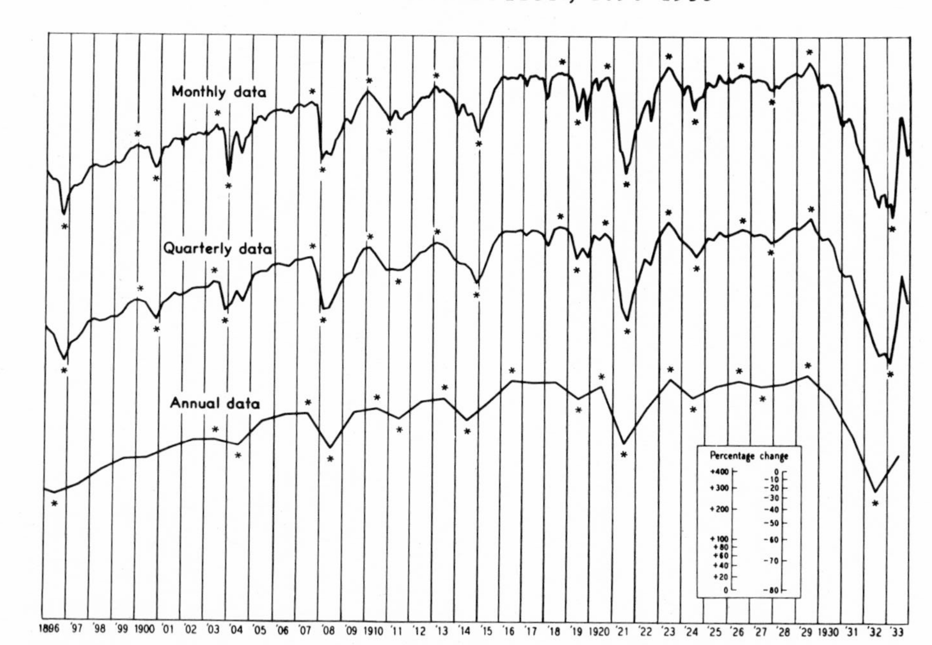

Note: The monthly and quarterly figures are seasonally adjusted. Asterisks identify peaks and troughs of specific cycles.

Source: Burns and Mitchell, *op. cit.*, p. 209.

ingful nine-stage patterns. In the five-stage pattern, partitioning is radically simplified. The two intermediate standings between peak and trough standings are based on weighted averages of the observations between peaks and troughs, or—for one-year phases—on the peak and trough observations themselves.[40]

Finally, the analysis of annual data does not provide B-4 tables, that is, tables with conformity measures after adjustment for average timing differences. The reason is that measures of leads and lags based on annual data are too crude to provide a basis for useful adjustments —a fact mentioned earlier in this section. Simple conformity measures, without adjustment for timing differences are, however, available in the cyclical analysis of annual data.

[40] *Ibid.*, p. 199. The substitution of five-stage patterns is optional since the user may wish to obtain nine-stage patterns, particularly in the analysis of long cycles. Incidentally, only a nine- or a five-stage pattern can be opted. If the user wants both, the analysis has to be repeated.

APPENDIX TO CHAPTER 3

A

SAMPLE RUN, BUSINESS CYCLE ANALYSIS, NONAGRICULTURAL EMPLOYMENT

Output Table 3A-1

EMPLOYEES IN NONAGRICULTURAL ESTABLISHMENTS, BLS

THOUSAND PERSONS

SAMPLE RUN

BASIC TIME SERIES

	JAN	FEB	MAR	APR	MAY	JUNE	JULY	AUG	SEPT	OCT	NOV	DEC
1929	32606	32595	32789	32885	32961	33057	33186	33348	33133	32971	32724	32294
1930	31918	31692	31423	31284	31069	30757	30434	30015	29725	29424	29134	28865
1931	28639	28456	28263	28220	28016	27747	27500	27166	26844	26478	26156	26048
1962	54695	55003	55162	55411	55502	55565	55657	55673	55767	55802	55874	55881
1963	55900	56044	56187	56368	56511	56601	56763	56768	56868	57070	57101	57291

DATA FOR 1929-38 RAISED TO THE LEVEL OF DATA FOR 1939-61

BY THE RATIO OF 1.0704

Output Table 3A-2

REFERENCE CYCLE ANALYSIS

CYCLE DATES			CYCLICAL AMPLITUDES			TROUGH TO TROUGH ANALYSIS				
			CYCLE RELATIVES			AMPLITUDES			AMPLITUDES PER	
TROUGH	PEAK	TROUGH	TROUGH	PEAK	TROUGH	RISE	FALL	TOTAL	RISE	FALL
1933 3	1937 5	1938 6	81.0	112.3	101.1	31.2	-11.2	42.4	.62	-.86
1938 6	1945 2	1945 10	77.4	112.5	103.9	35.1	-8.6	43.7	.44	-1.07
1945 10	1948 11	1949 10	89.1	104.1	99.8	15.0	-4.3	19.3	.41	-.39
1949 10	1953 7	1954 8	90.2	105.1	101.6	15.0	-3.5	18.4	.33	-.27
1954 8	1957 7	1958 4	94.3	102.7	98.7	8.3	-3.9	12.2	.24	-.44
1958 4	1960 5	1961 2	96.0	102.5	100.6	6.5	-1.8	8.3	.26	-.20
TOTAL			528.0	639.1	605.8	111.1	-33.3	144.4	2.29	-3.23
AVERAGE			88.0	106.5	101.0	18.5	-5.5	24.1	.38	-.54
AVERAGE DEVIATION			5.8	3.9	1.3	9.8	2.9	12.6	.11	.29
WEIGHTED AVERAGE									.41	-.53

Output Table 3A-3

CYCLE PATTERNS TROUGH TO TROUGH ANALYSIS

CYCLE DATES			AVERAGE IN ORIGINAL UNITS									
TROUGH	PEAK	TROUGH	TROUGH	II	III	IV	PEAK	VI	VII	VIII	TROUGH	
1927 11	1929 8	1933 3	.0	.0	.0	.0	33222.3	31347.4	27680.9	24031.7	23030.7	
1933 3	1937 5	1938 6	23030.7	25621.6	27580.0	30308.1	31903.7	32044.3	30813.5	29260.0	28725.0	2
1938 6	1945 2	1945 10	28725.0	30664.5	37473.8	42129.1	41740.0	41498.5	40886.7	39353.5	38559.3	3
1945 10	1948 11	1949 10	38559.3	40908.7	43657.9	44756.9	45077.3	44706.0	43936.5	43544.7	43215.0	4
1949 10	1953 7	1954 8	43215.0	45071.3	47978.4	49480.3	50377.7	50109.3	49371.0	48854.8	48710.7	4
1954 8	1957 7	1958 4	48710.7	49728.4	51826.8	52855.0	53004.3	52834.0	52428.5	51577.0	50975.7	5
1958 4	1960 5	1961 2	50975.7	51318.4	53154.1	53960.5	54414.3	54239.3	54028.5	53654.0	53465.3	5

Output Table 3A-4

EMPLOYEES IN NONAGRICULTURAL ESTABLISHMENTS, BLS 8268

THOUSAND PERSONS

EFERENCE CYCLE ANALYSIS

CYCLE PATTERNS TROUGH TO TROUGH ANALYSIS

CYCLE DATES			AVERAGE IN CYCLE RELATIVES									
ROUGH	PEAK	TROUGH	TROUGH	II	III	IV	PEAK	VI	VII	VIII	TROUGH	CYCLE
933 3	1937 5	1938 6	81.0	90.2	97.1	106.7	112.3	112.8	108.4	103.0	101.1	28411.3
938 6	1945 2	1945 10	77.4	82.7	101.0	113.6	112.5	111.9	110.2	106.1	103.9	37092.4
945 10	1948 11	1949 10	89.1	94.5	100.8	103.4	104.1	103.3	101.5	100.6	99.8	43294.0
949 10	1953 7	1954 8	90.2	94.0	100.1	103.2	105.1	104.5	103.0	101.9	101.6	47921.9
954 8	1957 7	1958 4	94.3	96.3	100.4	102.4	102.7	102.3	101.6	99.9	98.7	51618.9
958 4	1960 5	1961 2	96.0	96.6	100.1	101.6	102.5	102.1	101.7	101.0	100.6	53109.9
OTAL			528.0	554.2	599.5	630.8	639.1	636.9	626.4	612.4	605.8	
ERAGE			88.0	92.4	99.9	105.1	106.5	106.1	104.4	102.1	101.0	
ERAGE DEVIATION			5.8	4.0	1.0	3.3	3.9	4.1	3.3	1.6	1.3	

Output Table 3A-5

CYCLE PATTERNS TROUGH TO TROUGH ANALYSIS

CYCLE DATES			STAGE TO STAGE CHANGE OF CYCLE RELATIVES, TOTAL CHANGE							
ROUGH	PEAK	TROUGH	I-II	II-III	III-IV	IV-V	V-VI	VI-VII	VII-VIII	VIII-IX
933 3	1937 5	1938 6	9.1	6.9	9.6	5.6	.5	-4.3	-5.5	-1.9
938 6	1945 2	1945 10	5.2	18.4	12.5	-1.0	-.6	-1.7	-4.1	-2.1
945 10	1948 11	1949 10	5.4	6.4	2.5	.8	-.9	-1.8	-.9	-.8
949 10	1953 7	1954 8	3.9	6.1	3.1	1.9	-.6	-1.5	-1.1	-.3
954 8	1957 7	1958 4	2.0	4.1	2.0	.3	-.3	-.8	-1.7	-1.2
958 4	1960 5	1961 2	.7	3.5	1.5	.9	-.3	-.4	-.7	-.4
OTAL			26.2	45.3	31.3	8.3	-2.2	-10.5	-14.0	-6.6
ERAGE			4.4	7.5	5.2	1.4	-.4	-1.7	-2.3	-1.1
ERAGE DEVIATION			2.2	3.6	3.9	1.6	.3	.9	1.6	.6

Output Table 3A-6

CYCLE PATTERNS TROUGH TO TROUGH ANALYSIS

CYCLE DATES			STAGE TO STAGE CHANGE OF CYCLE RELATIVES CHANGE PER MONTH							
ROUGH	PEAK	TROUGH	I-II	II-III	III-IV	IV-V	V-VI	VI-VII	VII-VIII	VIII-IX
933 3	1937 5	1938 6	1.07	.42	.58	.66	.20	-1.08	-1.37	-.76
938 6	1945 2	1945 10	.38	.69	.47	-.08	-.43	-.66	-1.66	-1.43
945 10	1948 11	1949 10	.83	.53	.21	.12	-.43	-.51	-.26	-.38
949 10	1953 7	1954 8	.48	.42	.21	.23	-.22	-.39	-.27	-.12
954 8	1957 7	1958 4	.33	.36	.17	.04	-.16	-.31	-.66	-.59
958 4	1960 5	1961 2	.14	.43	.19	.19	-.17	-.16	-.29	-.18
OTAL			3.24	2.84	1.84	1.16	-1.21	-3.11	-4.50	-3.44
ERAGE			.54	.47	.31	.19	-.20	-.52	-.75	-.57
ERAGE DEVIATION			.27	.09	.14	.17	.16	.23	.51	.35
IGHTED AVERAGE			.56	.51	.35	.18	-.18	-.55	-.73	-.53

Output Table 3A-7

EMPLOYEES IN NONAGRICULTURAL ESTABLISHMENTS, BLS 82

THOUSAND PERSONS

REFERENCE CYCLE ANALYSIS

INTERVALS BETWEEN MIDPOINTS OF CYCLE STAGES IN MONTHS TROUGH TO TROUGH ANALYSIS

CYCLE DATES	I-II	II-III	III-IV	IV-V	V-VI	VI-VII	VII-VIII	VIII-IX
1933 3 1937 5 1938 6	8.5	16.5	16.5	8.5	2.5	4.0	4.0	2.5
1938 6 1945 2 1945 10	13.5	26.5	26.5	13.5	1.5	2.5	2.5	1.5
1945 10 1948 11 1949 10	6.5	12.0	12.0	6.5	2.0	3.5	3.5	2.0
1949 10 1953 7 1954 8	8.0	14.5	14.5	8.0	2.5	4.0	4.0	2.5
1954 8 1957 7 1958 4	6.0	11.5	11.5	6.0	2.0	2.5	2.5	2.0
1958 4 1960 5 1961 2	4.5	8.0	8.0	4.5	2.0	2.5	2.5	2.0
TOTAL	47.0	89.0	89.0	47.0	12.5	19.0	19.0	12.5
AV.	7.8	14.8	14.8	7.8	2.1	3.2	3.2	2.1

Output Table 3A-8

MEASURES OF SECULAR MOVEMENTS TROUGH TO TROUGH ANALYSIS

CYCLE DATES			AVERAGE MONTHLY STANDING			PERCENT CHANGE FROM PRECEDING PHASE		PERCENT CHANGE FROM PRECED CYCLE ON BASE OF			
								PRECEDING CYCLE		AVERAGE OF G AND PRECEDING C	
TROUGH	PEAK	TROUGH	EXP-N	CONT-N	FULL CYCLE	EXP-N	CONT-N	TOTAL	PER MO	TOTAL	PE
1927 11	1929 8	1933 3	.0	27695.8	.0	.0	.0	.0	.00	.0	
1933 3	1937 5	1938 6	27822.5	30675.5	28411.3	.5	10.3	.0	.00	.0	
1938 6	1945 2	1945 10	36745.5	40561.4	37092.4	19.8	10.4	30.6	.40	26.5	
1945 10	1948 11	1949 10	43072.0	44041.0	43294.0	6.2	2.2	16.7	.25	15.4	
1949 10	1953 7	1954 8	47479.6	49452.8	47921.9	7.8	4.2	10.7	.20	10.1	
1954 8	1957 7	1958 4	51462.3	52227.7	51618.9	4.1	1.5	7.7	.15	7.4	
1958 4	1960 5	1961 2	52804.7	53957.9	53109.9	1.1	2.2	2.9	.07	2.8	
TOTAL										62.4	
AVERAGE										12.5	
AVERAGE DEVIATION										6.8	
WEIGHTED AVERAGE											

Output Table 3A-9

DURATION OF CYCLICAL MOVEMENTS IN MONTHS

CYCLE DATES	EXPANSION	CONTRACTION	FULL CYCLE
1933 3 1937 5 1938 6	50	13	63
1938 6 1945 2 1945 10	80	8	88
1945 10 1948 11 1949 10	37	11	48
1949 10 1953 7 1954 8	45	13	58
1954 8 1957 7 1958 4	35	9	44
1958 4 1960 5 1961 2	25	9	34
TOTAL	272	63	335
AVERAGE	45.3	10.5	55.8
AVERAGE DEVIATION	13.1	1.8	13.8

Output Table 3A-10

EMPLOYEES IN NONAGRICULTURAL ESTABLISHMENTS, BLS 8268

THOUSAND PERSONS

FERENCE CYCLE ANALYSIS

CYCLE DATES						CYCLICAL AMPLITUDES CYCLE RELATIVES			PEAK TO PEAK ANALYSIS AMPLITUDES			AMPLITUDES PER MONTH		
AK		TROUGH		PEAK		PEAK	TROUGH	PEAK	FALL	RISE	TOTAL	FALL	RISE	TOTAL
29	8	1933	3	1937	5	119.6	82.9	114.9	-36.7	32.0	-68.7	-.85	.64	-.74
7	5	1938	6	1945	2	88.9	80.0	116.3	-8.9	36.3	-45.1	-.63	.45	-.48
5	2	1945	10	1948	11	97.9	90.4	105.7	-7.5	15.3	-22.8	-.93	.41	-.51
8	11	1949	10	1953	7	96.3	92.3	107.6	-4.0	15.3	-19.3	-.36	.34	-.34
3	7	1954	8	1957	7	98.9	95.6	104.1	-3.3	8.4	-11.7	-.25	.24	-.24
7	7	1958	4	1960	5	100.6	96.8	103.3	-3.8	6.5	-10.4	-.43	.26	-.30
TAL						602.3	538.1	651.9	-64.2	113.8	-178.0	-3.51	2.35	-2.62
RAGE						100.4	89.7	108.7	-10.7	19.0	-29.7	-.58	.39	-.44
RAGE DEVIATION						6.5	5.5	4.6	8.7	10.1	18.2	.24	.11	.14
GHTED AVERAGE												-.66	.42	-.48

Output Table 3A-11

CYCLE PATTERNS PEAK TO PEAK ANALYSIS

CYCLE DATES					AVERAGE IN CYCLE RELATIVES									
					PEAK	II	III	IV	TROUGH	VI	VII	VIII	PEAK	CYCLE
8	1933	3	1937	5	119.6	112.9	99.7	86.5	82.9	92.3	99.3	109.1	114.9	27763.9
5	1938	6	1945	2	88.9	89.3	85.8	81.5	80.0	85.4	104.4	117.3	116.3	35897.0
2	1945	10	1948	11	97.9	97.3	95.9	92.3	90.4	96.0	102.4	105.0	105.7	42625.7
11	1949	10	1953	7	96.3	95.5	93.9	93.0	92.3	96.3	102.5	105.7	107.6	46804.2
7	1954	8	1957	7	98.9	98.4	97.0	95.9	95.6	97.7	101.8	103.8	104.1	50918.1
7	1958	4	1960	5	100.6	100.3	99.6	97.9	96.8	97.5	100.9	102.5	103.3	52651.9
L					602.3	593.7	571.8	547.3	538.1	565.0	611.3	643.4	651.9	
GE					100.4	99.0	95.3	91.2	89.7	94.2	101.9	107.2	108.7	
GE DEVIATION					6.5	5.1	3.6	4.8	5.5	3.6	1.2	4.0	4.6	

Output Table 3A-12

CYCLE PATTERNS PEAK TO PEAK ANALYSIS

CYCLE DATES					STAGE TO STAGE CHANGE OF CYCLE RELATIVES, TOTAL CHANGE							
	TROUGH		PEAK		I-II	II-III	III-IV	IV-V	V-VI	VI-VII	VII-VIII	VIII-IX
8	1933	3	1937	5	-6.7	-13.2	-13.2	-3.6	9.3	7.1	9.8	5.8
5	1938	6	1945	2	.4	-3.4	-4.3	-1.5	5.4	19.0	13.0	-1.1
2	1945	10	1948	11	-.6	-1.4	-3.6	-1.9	5.5	6.5	2.6	.8
11	1949	10	1953	7	-.8	-1.6	-.8	-.7	4.0	6.2	3.2	1.9
7	1954	8	1957	7	-.5	-1.4	-1.0	-.3	2.0	4.1	2.0	.3
7	1958	4	1960	5	-.3	-.8	-1.6	-1.1	.7	3.5	1.5	.9
L					-8.6	-21.9	-24.5	-9.1	26.9	46.3	32.1	8.5
GE					-1.4	-3.7	-4.1	-1.5	4.5	7.7	5.3	1.4
GE DEVIATION					1.8	3.2	3.1	.8	2.3	3.7	4.0	1.6

Output Table 3A-13

EMPLOYEES IN NONAGRICULTURAL ESTABLISHMENTS, BLS

THOUSAND PERSONS

REFERENCE CYCLE ANALYSIS

CYCLE PATTERNS PEAK TO PEAK ANALYSIS

CYCLE DATES						STAGE TO STAGE CHANGE OF CYCLE RELATIVES CHANGE PER MONTH							
PEAK		TROUGH		PEAK		I-II	II-III	III-IV	IV-V	V-VI	VI-VII	VII-VIII	VIII-IX
1929	8	1933	3	1937	5	-.90	-.94	-.94	-.48	1.10	.43	.59	.68
1937	5	1938	6	1945	2	.16	-.86	-1.08	-.60	.40	.71	.49	-.08
1945	2	1945	10	1948	11	-.38	-.57	-1.44	-1.25	.85	.54	.21	.12
1948	11	1949	10	1953	7	-.40	-.47	-.24	-.36	.50	.43	.22	.24
1953	7	1954	8	1957	7	-.22	-.36	-.25	-.12	.34	.36	.17	.05
1957	7	1958	4	1960	5	-.16	-.30	-.65	-.57	.14	.44	.19	.19
TOTAL						-1.89	-3.51	-4.59	-3.38	3.32	2.90	1.88	1.19
AVERAGE						-.31	-.58	-.77	-.56	.55	.48	.31	.20
AVERAGE DEVIATION						.24	.21	.38	.24	.28	.09	.15	.17
WEIGHTED AVERAGE						-.47	-.72	-.80	-.51	.57	.52	.36	.18

Output Table 3A-14

INTERVALS BETWEEN MIDPOINTS OF CYCLE STAGES

IN MONTHS PEAK TO PEAK ANALYSIS

CYCLE DATES						I-II	II-III	III-IV	IV-V	V-VI	VI-VII	VII-VIII	VIII-IX
1929	8	1933	3	1937	5	7.5	14.0	14.0	7.5	8.5	16.5	16.5	8.
1937	5	1938	6	1945	2	2.5	4.0	4.0	2.5	13.5	26.5	26.5	13.
1945	2	1945	10	1948	11	1.5	2.5	2.5	1.5	6.5	12.0	12.0	6.
1948	11	1949	10	1953	7	2.0	3.5	3.5	2.0	8.0	14.5	14.5	8.
1953	7	1954	8	1957	7	2.5	4.0	4.0	2.5	6.0	11.5	11.5	6.
1957	7	1958	4	1960	5	2.0	2.5	2.5	2.0	4.5	8.0	8.0	4.
TOTAL						18.0	30.5	30.5	18.0	47.0	89.0	89.0	47.
AV.						3.0	5.1	5.1	3.0	7.8	14.8	14.8	7.

Output Table 3A-15

MEASURES OF SECULAR MOVEMENTS PEAK TO PEAK ANALYSIS

CYCLE DATES						AVERAGE MONTHLY STANDING			PERCENT CHANGE FROM PRECEDING PHASE		PERCENT CHANGE FROM PRECED CYCLE ON BASE OF			
											PRECEDING CYCLE		AVERAGE OF GI AND PRECEDING CY	
PEAK		TROUGH		PEAK		CONT-N	EXP-N	FULL CYCLE	CONT-N	EXP-N	TOTAL	PER MO	TOTAL	PER
1929	8	1933	3	1937	5	27695.8	27822.5	27763.9	.0	.5	.0	.00	.0	
1937	5	1938	6	1945	2	30675.5	36745.5	35897.0	10.3	19.8	29.3	.31	25.6	
1945	2	1945	10	1948	11	40561.4	43072.0	42625.7	10.4	6.2	18.7	.27	17.1	
1948	11	1949	10	1953	7	44041.0	47479.6	46804.2	2.2	7.8	9.8	.19	9.3	
1953	7	1954	8	1957	7	49452.8	51462.3	50918.1	4.2	4.1	8.8	.17	8.4	
1957	7	1958	4	1960	5	52227.7	52804.7	52651.9	1.5	1.1	3.4	.08	3.3	
1960	5	1961	2			53957.9	.0	.0	2.2	.0	.0	.00	.0	
TOTAL													63.8	
AVERAGE													12.8	
AVERAGE DEVIATION													6.9	
WEIGHTED AVERAGE														

Output Table 3A-16

EMPLOYEES IN NONAGRICULTURAL ESTABLISHMENTS, BLS

THOUSAND PERSONS

REFERENCE CYCLE ANALYSIS

DURATION OF CYCLICAL MOVEMENTS IN MONTHS

CYCLE DATES			CONTRACTION	EXPANSION	FULL CYCLE
1929 8	1933 3	1937 5	43	50	93
1937 5	1938 6	1945 2	13	80	93
1945 2	1945 10	1948 11	8	37	45
1948 11	1949 10	1953 7	11	45	56
1953 7	1954 8	1957 7	13	35	48
1957 7	1958 4	1960 5	9	25	34
TOTAL			97	272	369
AVERAGE			16.2	45.3	61.5
AVERAGE DEVIATION			8.9	13.1	21.0

Output Table 3A-17

CONFORMITY TO BUSINESS CYCLES

AVERAGE CHANGE PER MONTH, IN CYCLE RELATIVES

ON TROUGH-PEAK-TROUGH BASIS

CYCLE DATES JGH	PEAK	TROUGH	DURING EXPANSION	DURING CONT.N	CON.MINUS PRECED. EXP.
3	1937 5	1938 6	.62	-.86	-1.48
6	1945 2	1945 10	.44	-1.07	-1.51
10	1948 11	1949 10	.41	-.39	-.80
10	1953 7	1954 8	.33	-.27	-.60
8	1957 7	1958 4	.24	-.44	-.67
4	1960 5	1961 2	.26	-.20	-.46
AGE			.38	-.54	-.92
AGE DEVIATION			.11	.29	.38

ON PEAK-TROUGH-PEAK BASIS

CYCLE DATES PEAK	TROUGH	PEAK	DURING CONT.N	DURING EXPANSION	CON.MINUS SUCCED. EXP.
1929 8	1933 3	1937 5	-.85	.64	-1.49
1937 5	1938 6	1945 2	-.68	.45	-1.13
1945 2	1945 10	1948 11	-.93	.41	-1.35
1948 11	1949 10	1953 7	-.36	.34	-.70
1953 7	1954 8	1957 7	-.25	.24	-.49
1957 7	1958 4	1960 5	-.43	.26	-.69
			-.58	.39	-.97
			.24	.11	.35

) OF CONFORMITY TO REFERENCE	TPT DURING EXPANSION	TPT DURING CONT.N	TPT CON.MINUS PRECED. EXP.		PTP DURING CONT.N	PTP DURING EXPANSION	PTP CON.MINUS SUCCED. EXP.
NSIONS	100			100		100	
RACTIONS		100		100	100		
CYCLES, TROUGH TO TROUGH			100				
CYCLES, PEAK-TO-PEAK							100
CYCLES, BOTH WAYS				100			

Output Table 3A-18

EMPLOYEES IN NONAGRICULTURAL ESTABLISHMENTS, BLS 82

THOUSAND PERSONS

REFERENCE CYCLE ANALYSIS

-TIMING DIFFS RECOGNIZED BY SHIFTING REF. DATES,TROUGH MONTH, PEAK -1 MONTH

			CYCLICAL AMPLITUDES			TROUGH TO TROUGH ANALYSIS					
CYCLE DATES			CYCLE RELATIVES			AMPLITUDES			AMPLITUDES PER MONT		
TROUGH	PEAK	TROUGH	TROUGH	PEAK	TROUGH	RISE	FALL	TOTAL	RISE	FALL	TOTAL
1933 3	1937 4	1938 6	81.0	112.0	101.1	30.9	-10.9	41.8	.63	-.78	.66
1938 6	1945 1	1945 10	77.4	112.6	103.9	35.1	-8.6	43.8	.44	-.96	.5
1945 10	1948 10	1949 10	89.1	104.2	99.8	15.1	-4.4	19.5	.42	-.36	.4
1949 10	1953 6	1954 8	90.2	105.2	101.6	15.0	-3.5	18.6	.34	-.25	.3
1954 8	1957 6	1958 4	94.3	102.7	98.7	8.3	-3.9	12.3	.24	-.39	.2
1958 4	1960 4	1961 2	96.0	102.5	100.6	6.5	-1.9	8.4	.27	-.19	.2
TOTAL			528.0	639.1	605.8	111.1	-33.3	144.3	2.35	-2.94	2.4
AVERAGE			88.0	106.5	101.0	18.5	-5.5	24.1	.39	-.49	.4
AVERAGE DEVIATION			5.8	3.8	1.3	9.7	2.8	12.5	.11	.25	.1
WEIGHTED AVERAGE									.42	-.48	.4

Output Table 3A-19

-TIMING DIFFS RECOGNIZED BY SHIFTING REF. DATES,TROUGH MONTH, PEAK -1 MONTH

DURATION OF CYCLICAL MOVEMENTS IN MONTHS

CYCLE DATES			EXPANSION	CONTRACTION	FULL CYCLE
1933 3	1937 4	1938 6	49	14	63
1938 6	1945 1	1945 10	79	9	88
1945 10	1948 10	1949 10	36	12	48
1949 10	1953 6	1954 8	44	14	58
1954 8	1957 6	1958 4	34	10	44
1958 4	1960 4	1961 2	24	10	34
TOTAL			266	69	335
AVERAGE			44.3	11.5	55.8
AVERAGE DEVIATION			13.1	1.8	13.8

Output Table 3A-20

EMPLOYEES IN NONAGRICULTURAL ESTABLISHMENTS, BLS 8268

THOUSAND PERSONS

FERENCE CYCLE ANALYSIS

TIMING DIFFS RECOGNIZED BY SHIFTING REF. DATES,TROUGH MONTH, PEAK -1 MONTH

CYCLE DATES AK		TROUGH		PEAK		CYCLICAL AMPLITUDES CYCLE RELATIVES PEAK	TROUGH	PEAK	PEAK TO PEAK ANALYSIS AMPLITUDES FALL	RISE	TOTAL	AMPLITUDES PER MONTH FALL	RISE	TOTAL
29	7	1933	3	1937	4	119.5	82.9	114.5	-36.6	31.6	-68.2	-.83	.65	-.73
37	4	1938	6	1945	1	88.9	80.2	116.7	-8.7	36.4	-45.1	-.62	.46	-.48
45	1	1945	10	1948	10	98.1	90.6	106.0	-7.5	15.4	-23.0	-.84	.43	-.51
48	10	1949	10	1953	6	96.6	92.5	107.9	-4.1	15.4	-19.5	-.34	.35	-.35
53	6	1954	8	1957	6	99.1	95.8	104.2	-3.3	8.4	-11.8	-.24	.25	-.24
57	6	1958	4	1960	4	100.7	96.9	103.5	-3.8	6.6	-10.4	-.38	.27	-.30
)TAL						602.9	538.9	652.8	-64.0	113.9	-177.9	-3.24	2.41	-2.62
ERAGE						100.5	89.8	108.8	-10.7	19.0	-29.7	-.54	.40	-.44
ERAGE DEVIATION						6.4	5.5	4.5	8.6	10.0	18.0	.22	.11	.14
GHTED AVERAGE												-.62	.43	-.48

Output Table 3A-21

CONFORMITY TO BUSINESS CYCLES

AVERAGE CHANGE PER MONTH, IN CYCLE RELATIVES

ON TROUGH-PEAK-TROUGH BASIS CYCLE DATES JGH		PEAK		TROUGH		DURING EXPANSION	DURING CONT.N	CON.MINUS PRECED. EXP.	
3	3	1937	4	1938	6	.63	-.78	-1.41	
3	6	1945	1	1945	10	.44	-.96	-1.40	
5	10	1948	10	1949	10	.42	-.36	-.78	
	10	1953	6	1954	8	.34	-.25	-.59	
	8	1957	6	1958	4	.24	-.39	-.64	
	4	1960	4	1961	2	.27	-.19	-.46	
AGE						.39	-.49	-.88	
AGE DEVIATION						.11	.25	.35	
O OF CONFORMITY TO REFERENCE									
NSIONS						100			100
RACTIONS							100		100
CYCLES, TROUGH TO TROUGH								100	
CYCLES,PEAK-TO-PEAK									
CYCLES,BOTH WAYS									100

ON PEAK-TROUGH-PEAK BASIS CYCLE DATES PEAK		TROUGH		PEAK		DURING CONT.N	DURING EXPANSION	CON.MINUS SUCCED. EXP.
1929	7	1933	3	1937	4	-.83	.65	-1.48
1937	4	1938	6	1945	1	-.62	.46	-1.08
1945	1	1945	10	1948	10	-.84	.43	-1.27
1948	10	1949	10	1953	6	-.34	.35	-.69
1953	6	1954	8	1957	6	-.24	.25	-.48
1957	6	1958	4	1960	4	-.38	.27	-.66
AGE						-.54	.40	-.94
AGE DEVIATION						.22	.11	.33
O OF CONFORMITY TO REFERENCE								
NSIONS							100	
RACTIONS						100		
CYCLES, TROUGH TO TROUGH								
CYCLES,PEAK-TO-PEAK								100
CYCLES,BOTH WAYS								

Output Table 3A-22

EMPLOYEES IN NONAGRICULTURAL ESTABLISHMENTS, BLS 82

THOUSAND PERSONS

SPECIFIC CYCLE ANALYSIS

CYCLICAL AMPLITUDES TROUGH TO TROUGH ANALYSIS

CYCLE DATES						CYCLE RELATIVES			AMPLITUDES			AMPLITUDES PER MONT		
TROUGH		PEAK		TROUGH		TROUGH	PEAK	TROUGH	RISE	FALL	TOTAL	RISE	FALL	TOTAL
1933	3	1937	7	1938	6	81.0	112.8	101.1	31.8	-11.8	43.6	.61	-1.07	.69
1938	6	1943	11	1945	9	77.5	115.0	105.3	37.5	-9.7	47.2	.58	-.44	.54
1945	9	1948	7	1949	10	90.4	104.2	100.0	13.8	-4.2	18.0	.41	-.28	.37
1949	10	1953	7	1954	8	90.2	105.1	101.6	15.0	-3.5	18.4	.33	-.27	.3
1954	8	1957	3	1958	5	94.4	102.8	98.6	8.5	-4.3	12.8	.27	-.31	.2
1958	5	1960	4	1961	2	95.6	102.4	100.5	6.7	-1.8	8.6	.29	-.18	.2
TOTAL						529.1	642.4	607.2	113.3	-35.2	148.5	2.49	-2.55	2.4
AVERAGE						88.2	107.1	101.2	18.9	-5.9	24.8	.41	-.42	.4
AVERAGE DEVIATION						6.0	4.6	1.5	10.5	3.2	13.8	.12	.22	.1
WEIGHTED AVERAGE												.45	-.41	.4

Output Table 3A-23

CYCLE PATTERNS TROUGH TO TROUGH ANALYSIS

AVERAGE IN ORIGINAL UNITS

CYCLE DATES															
TROUGH		PEAK		TROUGH		TROUGH	II	III	IV	PEAK	VI	VII	VIII	TROUGH	CYC
		1929	8	1933	3	.0	.0	.0	.0	33222.3	31347.4	27680.9	24031.7	23030.7	
1933	3	1937	7	1938	6	23030.7	25694.7	27692.6	30593.8	32068.3	31925.0	30300.3	29133.7	28725.0	28411
1938	6	1943	11	1945	9	28725.0	30361.4	35008.5	41295.1	42642.0	42159.3	41664.4	41108.7	39059.0	37077
1945	9	1948	7	1949	10	39059.0	40324.4	43410.5	44443.1	45029.7	45083.0	44334.8	43621.6	43215.0	4319
1949	10	1953	7	1954	8	43215.0	45071.3	47978.4	49480.3	50377.7	50109.3	49371.0	48854.8	48710.7	4792
1954	8	1957	3	1958	5	48710.7	49615.8	51604.7	52592.7	53079.3	53023.8	52671.8	51411.8	50867.7	5160
1958	5	1960	4	1961	2	50867.7	51389.3	53154.1	53875.6	54444.3	54302.7	54085.3	53654.0	53465.3	5317

Output Table 3A-24

CYCLE PATTERNS TROUGH TO TROUGH ANALYSIS

AVERAGE IN CYCLE RELATIVES

CYCLE DATES															
TROUGH		PEAK		TROUGH		TROUGH	II	III	IV	PEAK	VI	VII	VIII	TROUGH	CY
1933	3	1937	7	1938	6	81.0	90.4	97.5	107.7	112.8	112.4	106.6	102.5	101.1	2841
1938	6	1943	11	1945	9	77.5	81.9	94.4	111.4	115.0	113.7	112.4	110.8	105.3	3707
1945	9	1948	7	1949	10	90.4	93.3	100.5	102.9	104.2	104.4	102.6	101.0	100.0	431
1949	10	1953	7	1954	8	90.2	94.0	100.1	103.2	105.1	104.5	103.0	101.9	101.6	479
1954	8	1957	3	1958	5	94.4	96.1	100.0	101.9	102.8	102.8	102.1	99.6	98.6	516
1958	5	1960	4	1961	2	95.6	96.6	99.9	101.3	102.4	102.1	101.7	100.9	100.5	531
TOTAL						529.1	552.4	592.4	628.3	642.4	639.8	628.4	616.8	607.2	
AVERAGE						88.2	92.1	98.7	104.7	107.1	106.6	104.7	102.8	101.2	
AVERAGE DEVIATION						6.0	4.0	1.9	3.2	4.6	4.3	3.2	2.7	1.5	

Output Table 3A-25

EMPLOYEES IN NONAGRICULTURAL ESTABLISHMENTS, BLS 8268

THOUSAND PERSONS

PECIFIC CYCLE ANALYSIS

CYCLE PATTERNS TROUGH TO TROUGH ANALYSIS

CYCLE DATES						STAGE TO STAGE CHANGE OF CYCLE RELATIVES, TOTAL CHANGE							
ROUGH		PEAK		TROUGH		I-II	II-III	III-IV	IV-V	V-VI	VI-VII	VII-VIII	VIII-IX
933	3	1937	7	1938	6	9.4	7.0	10.2	5.2	-.5	-5.7	-4.1	-1.5
938	6	1943	11	1945	9	4.4	12.5	17.0	3.6	-1.3	-1.3	-1.5	-5.5
945	9	1948	7	1949	10	2.9	7.2	2.4	1.3	.1	-1.7	-1.7	-.9
949	10	1953	7	1954	8	3.9	6.1	3.1	1.9	-.6	-1.5	-1.1	-.3
954	8	1957	3	1958	5	1.8	3.9	1.9	.9	-.1	-.7	-2.4	-1.1
58	5	1960	4	1961	2	1.0	3.3	1.4	1.1	-.3	-.4	-.8	-.4
OTAL						23.3	40.0	36.0	14.0	-2.6	-11.4	-11.6	-9.6
ERAGE						3.9	6.7	6.0	2.3	-.4	-1.9	-1.9	-1.6
ERAGE DEVIATION						2.0	2.2	5.1	1.4	.3	1.3	.9	1.3

Output Table 3A-26

CYCLE PATTERNS TROUGH TO TROUGH ANALYSIS

CYCLE DATES						STAGE TO STAGE CHANGE OF CYCLE RELATIVES CHANGE PER MONTH							
UGH		PEAK		TROUGH		I-II	II-III	III-IV	IV-V	V-VI	VI-VII	VII-VIII	VIII-IX
33	3	1937	7	1938	6	1.04	.41	.60	.57	-.24	-1.63	-1.17	-.73
8	6	1943	11	1945	9	.40	.58	.79	.33	-.33	-.19	-.21	-1.38
5	9	1948	7	1949	10	.49	.65	.22	.22	.05	-.38	-.37	-.31
9	10	1953	7	1954	8	.48	.42	.21	.23	-.22	-.39	-.27	-.12
4	8	1957	3	1958	5	.32	.38	.19	.17	-.04	-.15	-.54	-.42
8	5	1960	4	1961	2	.25	.44	.18	.27	-.13	-.14	-.27	-.18
AL						2.98	2.89	2.19	1.80	-.92	-2.89	-2.83	-3.14
RAGE						.50	.48	.36	.30	-.15	-.48	-.47	-.52
RAGE DEVIATION						.18	.09	.22	.10	.11	.38	.26	.35
GHTED AVERAGE						.54	.49	.44	.32	-.16	-.43	-.44	-.60

Output Table 3A-27

INTERVALS BETWEEN MIDPOINTS OF CYCLE STAGES

IN MONTHS TROUGH TO TROUGH ANALYSIS

CYCLE DATES						I-II	II-III	III-IV	IV-V	V-VI	VI-VII	VII-VIII	VIII-IX
	3	1937	7	1938	6	9.0	17.0	17.0	9.0	2.0	3.5	3.5	2.0
	6	1943	11	1945	9	11.0	21.5	21.5	11.0	4.0	7.0	7.0	4.0
	9	1948	7	1949	10	6.0	11.0	11.0	6.0	3.0	4.5	4.5	3.0
	10	1953	7	1954	8	8.0	14.5	14.5	8.0	2.5	4.0	4.0	2.5
	8	1957	3	1958	5	5.5	10.0	10.0	5.5	2.5	4.5	4.5	2.5
	5	1960	4	1961	2	4.0	7.5	7.5	4.0	2.0	3.0	3.0	2.0
AL						43.5	81.5	81.5	43.5	16.0	26.5	26.5	16.0
						7.3	13.6	13.6	7.3	2.7	4.4	4.4	2.7

Output Table 3A-28

EMPLOYEES IN NONAGRICULTURAL ESTABLISHMENTS, BLS 826

THOUSAND PERSONS

SPECIFIC CYCLE ANALYSIS

MEASURES OF SECULAR MOVEMENTS TROUGH TO TROUGH ANALYSIS

CYCLE DATES			AVERAGE MONTHLY STANDING			PERCENT CHANGE FROM PRECEDING PHASE		PERCENT CHANGE FROM PRECEDING CYCLE ON BASE OF			
								PRECEDING CYCLE		AVERAGE OF GIVEN AND PRECEDING CYCLE	
TROUGH	PEAK	TROUGH	EXP-N	CONT-N	FULL CYCLE	EXP-N	CONT-N	TOTAL	PER MO	TOTAL	PER MO
	1929 8	1933 3	.0	27695.8	.0	.0	.0	.0	.00	.0	.00
1933 3	1937 7	1938 6	27983.6	30433.0	28411.3	1.0	8.8	.0	.00	.0	.00
1938 6	1943 11	1945 9	35548.4	41593.8	37077.1	16.8	17.0	30.5	.41	26.5	.35
1945 9	1948 7	1949 10	42697.5	44322.0	43194.8	2.7	3.8	16.5	.24	15.2	.22
1949 10	1953 7	1954 8	47479.6	49452.8	47921.9	7.1	4.2	10.9	.20	10.4	.19
1954 8	1957 3	1958 5	51259.0	52362.1	51602.2	3.7	2.2	7.7	.15	7.4	.14
1958 5	1960 4	1961 2	52816.3	54009.3	53177.8	.9	2.3	3.1	.08	3.0	.08
TOTAL										62.5	.9
AVERAGE										12.5	.2
AVERAGE DEVIATION										6.7	.0
WEIGHTED AVERAGE											.21

Output Table 3A-29

DURATION OF CYCLICAL MOVEMENTS IN MONTHS

CYCLE DATES			EXPANSION	CONTRACTION	FULL CYCLE
1933 3	1937 7	1938 6	52	11	63
1938 6	1943 11	1945 9	65	22	87
1945 9	1948 7	1949 10	34	15	49
1949 10	1953 7	1954 8	45	13	58
1954 8	1957 3	1958 5	31	14	45
1958 5	1960 4	1961 2	23	10	33
TOTAL			250	85	335
AVERAGE			41.7	14.2	55.8
AVERAGE DEVIATION			12.3	2.9	13.5

Output Table 3A-30

CYCLICAL AMPLITUDES PEAK TO PEAK ANALYSIS

CYCLE DATES			CYCLE RELATIVES			AMPLITUDES			AMPLITUDES PER M		
PEAK	TROUGH	PEAK	PEAK	TROUGH	PEAK	FALL	RISE	TOTAL	FALL	RISE	TO
1929 8	1933 3	1937 7	119.3	82.7	115.1	-36.6	32.4	-69.0	-.85	.62	-
1937 7	1938 6	1943 11	92.1	82.5	122.5	-9.6	40.0	-49.6	-.87	.62	-
1943 11	1945 9	1948 7	100.9	92.4	106.5	-8.5	14.1	-22.6	-.38	.42	-
1948 7	1949 10	1953 7	96.4	92.5	107.9	-3.9	15.3	-19.2	-.26	.34	-
1953 7	1954 8	1957 3	99.3	96.0	104.6	-3.3	8.6	-11.9	-.25	.28	-
1957 3	1958 5	1960 4	100.8	96.6	103.4	-4.2	6.8	-11.0	-.30	.29	-
TOTAL			608.8	542.7	660.0	-66.1	117.3	-183.4	-2.91	2.57	-2
AVERAGE			101.5	90.5	110.0	-11.0	19.5	-30.6	-.49	.43	-
AVERAGE DEVIATION			5.9	5.2	5.9	8.5	11.1	19.2	.25	.13	
WEIGHTED AVERAGE									-.56	.47	-

Output Table 3A-31

EMPLOYEES IN NONAGRICULTURAL ESTABLISHMENTS, BLS 8268

THOUSAND PERSONS

SPECIFIC CYCLE ANALYSIS

CYCLE PATTERNS PEAK TO PEAK ANALYSIS

CYCLE DATES						AVERAGE IN CYCLE RELATIVES									
						PEAK	II	III	IV	TROUGH	VI	VII	VIII	PEAK	CYCLE
1929	8	1933	3	1937	7	119.3	112.5	99.4	86.3	82.7	92.3	99.4	109.8	115.1	27853.3
937	7	1938	6	1943	11	92.1	91.7	87.0	83.7	82.5	87.2	100.6	118.6	122.5	34808.0
943	11	1945	9	1948	7	100.9	99.7	98.6	97.3	92.4	95.4	102.7	105.1	106.5	42263.9
948	7	1949	10	1953	7	96.4	96.5	94.9	93.4	92.5	96.5	102.8	106.0	107.9	46690.2
953	7	1954	8	1957	3	99.3	98.8	97.3	96.3	96.0	97.8	101.7	103.7	104.6	50725.4
957	3	1958	5	1960	4	100.8	100.7	100.0	97.7	96.6	97.6	101.0	102.3	103.4	52644.5
TOTAL						608.8	600.0	577.3	554.6	542.7	566.8	608.1	645.5	660.0	
VERAGE						101.5	100.0	96.2	92.4	90.5	94.5	101.3	107.6	110.0	
VERAGE DEVIATION						5.9	4.4	3.5	5.0	5.2	3.2	1.0	4.4	5.9	

Output Table 3A-32

CYCLE PATTERNS PEAK TO PEAK ANALYSIS

CYCLE DATES						STAGE TO STAGE CHANGE OF CYCLE RELATIVES, TOTAL CHANGE							
EAK		TROUGH		PEAK		I-II	II-III	III-IV	IV-V	V-VI	VI-VII	VII-VIII	VIII-IX
29	8	1933	3	1937	7	-6.7	-13.2	-13.1	-3.6	9.6	7.2	10.4	5.3
37	7	1938	6	1943	11	-.4	-4.7	-3.3	-1.2	4.7	13.3	18.0	3.9
43	11	1945	9	1948	7	-1.2	-1.2	-1.3	-4.9	3.0	7.3	2.5	1.4
48	7	1949	10	1953	7	.1	-1.6	-1.5	-.9	4.0	6.2	3.2	1.9
53	7	1954	8	1957	3	-.5	-1.4	-1.0	-.3	1.8	3.9	2.0	1.0
57	3	1958	5	1960	4	-.1	-.7	-2.4	-1.0	1.0	3.4	1.4	1.1
OTAL						-8.8	-22.7	-22.7	-11.9	24.1	41.3	37.5	14.5
ERAGE						-1.5	-3.8	-3.8	-2.0	4.0	6.9	6.2	2.4
ERAGE DEVIATION						1.8	3.4	3.1	1.5	2.1	2.4	5.3	1.4

Output Table 3A-33

CYCLE PATTERNS PEAK TO PEAK ANALYSIS

CYCLE DATES						STAGE TO STAGE CHANGE OF CYCLE RELATIVES CHANGE PER MONTH							
AK		TROUGH		PEAK		I-II	II-III	III-IV	IV-V	V-VI	VI-VII	VII-VIII	VIII-IX
29	8	1933	3	1937	7	-.90	-.94	-.94	-.48	1.06	.42	.61	.59
7	7	1938	6	1943	11	-.20	-1.33	-.96	-.59	.43	.62	.84	.35
3	11	1945	9	1948	7	-.29	-.17	-.18	-1.22	.50	.66	.22	.23
8	7	1949	10	1953	7	.04	-.36	-.34	-.30	.50	.43	.22	.24
3	7	1954	8	1957	3	-.21	-.36	-.26	-.12	.33	.39	.20	.17
7	3	1958	5	1960	4	-.04	-.14	-.53	-.42	.25	.45	.18	.27
AL						-1.60	-3.30	-3.20	-3.12	3.06	2.97	2.27	1.85
RAGE						-.27	-.55	-.53	-.52	.51	.49	.38	.31
RAGE DEVIATION						.22	.39	.27	.25	.18	.10	.23	.11
GHTED AVERAGE						-.41	-.60	-.60	-.55	.55	.51	.46	.33

Output Table 3A-34

EMPLOYEES IN NONAGRICULTURAL ESTABLISHMENTS, BLS 826

THOUSAND PERSONS

SPECIFIC CYCLE ANALYSIS

INTERVALS BETWEEN MIDPOINTS OF CYCLE STAGES IN MONTHS PEAK TO PEAK ANALYSIS

CYCLE DATES						I-II	II-III	III-IV	IV-V	V-VI	VI-VII	VII-VIII	VIII-IX
1929	8	1933	3	1937	7	7.5	14.0	14.0	7.5	9.0	17.0	17.0	9.0
1937	7	1938	6	1943	11	2.0	3.5	3.5	2.0	11.0	21.5	21.5	11.0
1943	11	1945	9	1948	7	4.0	7.0	7.0	4.0	6.0	11.0	11.0	6.0
1948	7	1949	10	1953	7	3.0	4.5	4.5	3.0	8.0	14.5	14.5	8.0
1953	7	1954	8	1957	3	2.5	4.0	4.0	2.5	5.5	10.0	10.0	5.5
1957	3	1958	5	1960	4	2.5	4.5	4.5	2.5	4.0	7.5	7.5	4.0
TOTAL						21.5	37.5	37.5	21.5	43.5	81.5	81.5	43.5
AV.						3.6	6.3	6.3	3.6	7.3	13.6	13.6	7.3

Output Table 3A-35

CYCLE DATES						MEASURES OF SECULAR MOVEMENTS AVERAGE MONTHLY STANDING			PEAK TO PEAK ANALYSIS PERCENT CHANGE FROM PRECEDING PHASE		PERCENT CHANGE FROM PRECEDIN CYCLF ON BASE OF			
											PRECEDING CYCLE		AVERAGE OF GIVE AND PRECEDING CYCL	
PEAK		TROUGH		PEAK		CONT-N	EXP-N	FULL CYCLE	CONT-N	EXP-N	TOTAL	PER MO	TOTAL	PER M
1929	8	1933	3	1937	7	27695.8	27983.6	27853.3	.0	1.0	.0	.00	.0	.[illegible]
1937	7	1938	6	1943	11	30433.0	35548.4	34808.0	8.8	16.8	25.0	.29	22.2	.[illegible]
1943	11	1945	9	1948	7	41593.8	42697.5	42263.9	17.0	2.7	21.4	.32	19.3	.[illegible]
1948	7	1949	10	1953	7	44322.0	47479.6	46690.2	3.8	7.1	10.5	.18	10.0	.[illegible]
1953	7	1954	8	1957	3	49452.8	51259.0	50725.4	4.2	3.7	8.6	.17	8.3	.[illegible]
1957	3	1958	5	1960	4	52362.1	52816.3	52644.5	2.2	.9	3.8	.09	3.7	.[illegible]
1960	4	1961	2			54009.3	.0	.0	2.3	.0	.0	.00	.0	.[illegible]
TOTAL													63.5	.[illegible]
AVERAGE													12.7	.[illegible]
AVERAGE DEVIATION													6.5	.[illegible]
WEIGHTED AVERAGE														.2[illegible]

Output Table 3A-36

DURATION OF CYCLICAL MOVEMENTS IN MONTHS

CYCLE DATES						CONTRACTION	EXPANSION	FULL CYCLE
1929	8	1933	3	1937	7	43	52	95
1937	7	1938	6	1943	11	11	65	76
1943	11	1945	9	1948	7	22	34	56
1948	7	1949	10	1953	7	15	45	60
1953	7	1954	8	1957	3	13	31	44
1957	3	1958	5	1960	4	14	23	37
TOTAL						118	250	368
AVERAGE						19.7	41.7	61.3
AVERAGE DEVIATION						8.6	12.3	16.1

APPENDIX TO CHAPTER 3

B

SAMPLE RUN, BUSINESS CYCLE ANALYSIS, UNEMPLOYMENT RATE

Output Table 3B-1

UNEMPLOYED RATE, NICB, CENSUS (INVERTED) 829

PER CENT X 100

SAMPLE RUN

BUSINESS CYCLE ANALYSIS, ABSOLUTE CHANGES

BASIC TIME SERIES

	JAN	FEB	MAR	APR	MAY	JUNE	JULY	AUG	SEPT	OCT	NOV	DEC	TOTAL
1929				76	181	227	87	4	100	254	208	229	136
1930	243	343	399	396	373	415	552	730	857	991	1185	1308	779
1931	1229	1283	1295	1313	1353	1432	1522	1651	1741	1842	1974	2100	1873
1962	580	553	553	561	551	549	543	574	564	543	583	549	67
1963	573	595	565	567	589	566	561	554	554	560	591	549	68

Output Table 3B-2

REFERENCE CYCLE ANALYSIS

CYCLICAL AMPLITUDES POSITIVE PLAN

CYCLE DATES			CYCLE DEVIATIONS			AMPLITUDES			AMPLITUDES PER MONTH		
TROUGH	PEAK	TROUGH	TROUGH	PEAK	TROUGH	RISE	FALL	TOTAL	RISE	FALL	TOTAL
1933 3	1937 5	1938 6	862.1	-615.2	291.5	-1477.3	906.7	-2384.0	-29.55	69.74	-37.
1938 6	1945 2	1945 10	1276.8	-779.6	-521.6	-2056.3	258.0	-2314.3	-25.70	32.25	-26.
1945 10	1948 11	1949 10	-61.2	-41.2	262.2	20.0	303.3	-283.3	.54	27.58	-5.
1949 10	1953 7	1954 8	295.7	-141.0	189.3	-436.7	330.3	-767.0	-9.70	25.41	-13.
1954 8	1957 7	1958 4	131.9	-38.5	254.5	-170.3	293.0	-463.3	-4.87	32.55	-10.
1958 4	1960 5	1961 2	119.6	-69.4	83.3	-189.0	152.7	-341.7	-7.56	16.96	-10.
TOTAL			2624.9	-1684.8	559.2	-4309.7	2244.0	-6553.7	-76.84	204.50	-103.
AVERAGE			437.5	-280.8	93.2	-718.3	374.0	-1092.3	-12.81	34.08	-17.
AVERAGE DEVIATION			421.3	277.7	208.2	699.0	177.6	837.9	9.88	11.89	9.
WEIGHTED AVERAGE									-15.84	35.62	-19.

Output Table 3B-3

CYCLE PATTERNS POSITIVE PLAN

STANDINGS, IN ORIGINAL UNITS

CYCLE DATES												
TROUGH	PEAK	TROUGH	TROUGH	II	III	IV	PEAK	VI	VII	VIII	TROUGH	CYCL
1927 11	1929 8	1933 3	.0	.0	.0	.0	63.7	435.0	1516.3	2516.9	2745.0	
1933 3	1937 5	1938 6	2745.0	2275.6	2045.0	1502.1	1267.7	1243.8	1613.5	2063.8	2174.3	188
1938 6	1945 2	1945 10	2174.3	1801.8	925.6	174.2	118.0	115.0	148.7	270.0	376.0	89
1945 10	1948 11	1949 10	376.0	430.6	405.2	371.4	396.0	445.7	555.3	655.7	699.3	43
1949 10	1953 7	1954 8	699.3	533.3	325.5	291.2	262.7	303.8	509.0	571.0	593.0	40
1954 8	1957 7	1958 4	593.0	490.1	423.3	408.7	422.7	450.7	498.0	629.3	715.7	46
1958 4	1960 5	1961 2	715.7	699.0	542.4	540.0	526.7	554.0	586.5	650.0	679.3	59

Output Table 3B-4

UNEMPLOYED RATE, NICB, CENSUS (INVERTED) 8292

PER CENT X 100

FERENCE CYCLE ANALYSIS

CYCLE PATTERNS POSITIVE PLAN
STANDINGS, DEVIATIONS FROM CYCLE BASE

CYCLE DATES															
ROUGH		PEAK		TROUGH		TROUGH	II	III	IV	PEAK	VI	VII	VIII	TROUGH	CYCLE
33	3	1937	5	1938	6	862.1	392.8	162.1	-380.8	-615.2	-639.1	-269.4	180.9	291.5	1882.9
38	6	1945	2	1945	10	1276.8	904.3	28.0	-723.4	-779.6	-782.6	-748.9	-627.6	-521.6	897.6
45	10	1948	11	1949	10	-61.2	-6.6	-32.0	-65.7	-41.2	8.5	118.1	218.5	262.2	437.2
49	10	1953	7	1954	8	295.7	129.7	-78.2	-112.5	-141.0	-99.9	105.3	167.3	189.3	403.7
54	8	1957	7	1958	4	131.9	29.0	-37.9	-52.4	-38.5	-10.5	36.9	168.2	254.5	461.1
58	4	1960	5	1961	2	119.6	103.0	-53.7	-56.0	-69.4	-42.0	-9.5	54.0	83.3	596.0
TAL						2624.9	1552.1	-11.6	-1390.9	-1684.8	-1565.6	-767.5	161.3	559.2	
RAGE						437.5	258.7	-1.9	-231.8	-280.8	-260.9	-127.9	26.9	93.2	
RAGE DEVIATION						421.3	259.9	64.7	213.5	277.7	299.9	254.1	218.2	208.2	

Output Table 3B-5

CYCLE PATTERNS POSITIVE PLAN

STAGE TO STAGE CHANGE OF STANDINGS, TOTAL CHANGE

CYCLE DATES													
ROUGH		PEAK		TROUGH		I-II	II-III	III-IV	IV-V	V-VI	VI-VII	VII-VIII	VIII-IX
1933	3	1937	5	1938	6	-469.4	-230.6	-542.9	-234.4	-23.9	369.8	450.3	110.6
938	6	1945	2	1945	10	-372.5	-876.3	-751.4	-56.2	-3.0	33.7	121.3	106.0
945	10	1948	11	1949	10	54.6	-25.4	-33.8	24.6	49.7	109.6	100.4	43.7
949	10	1953	7	1954	8	-166.0	-207.8	-34.3	-28.5	41.1	205.3	62.0	22.0
954	8	1957	7	1958	4	-102.9	-66.8	-14.5	13.9	28.0	47.3	131.3	86.3
958	4	1960	5	1961	2	-16.7	-156.6	-2.4	-13.3	27.3	32.5	63.5	29.3
TOTAL						-1072.9	-1563.6	-1379.3	-293.9	119.2	798.1	928.8	397.9
VERAGE						-178.8	-260.6	-229.9	-49.0	19.9	133.0	154.8	66.3
VERAGE DEVIATION						161.4	205.2	278.2	64.2	22.2	103.0	98.5	34.6

Output Table 3B-6

UNEMPLOYED RATE, NICB, CENSUS (INVERTED) 82

PER CENT X 100

REFERENCE CYCLE ANALYSIS

CYCLE PATTERNS POSITIVE PLAN

CYCLE DATES			STAGE TO STAGE CHANGE OF STANDINGS, CHANGE PER MONTH							
TROUGH	PEAK	TROUGH	I-II	II-III	III-IV	IV-V	V-VI	VI-VII	VII-VIII	VIII-IX
1933 3	1937 5	1938 6	-55.2	-14.0	-32.9	-27.6	-9.6	92.4	112.6	44.2
1938 6	1945 2	1945 10	-27.6	-33.1	-28.4	-7.1	-.0	13.5	48.5	70.7
1945 10	1948 11	1949 10	8.4	-2.7	.0	2.5	24.8	31.3	28.7	21.8
1949 10	1953 7	1954 8	-20.8	-14.3	-2.8	.0	16.4	51.3	15.5	8.8
1954 8	1957 7	1958 4	-17.1	.0	-1.7	1.3	14.0	18.9	52.5	43.2
1958 4	1960 5	1961 2	.0	-19.6	.0	-1.3	13.7	13.0	25.4	14.7
TOTAL			-112.3	-83.6	-65.7	-32.2	59.4	220.4	283.2	203.4
AVERAGE			-18.7	-13.9	-11.0	-5.4	9.9	36.7	47.2	33.9
AVERAGE DEVIATION			15.8	8.4	13.1	8.0	9.8	23.4	24.0	18.8
WEIGHTED AVERAGE			-22.8	-17.6	-15.5	-6.3	9.5	42.0	48.9	31.8

Output Table 3B-7

INTERVALS BETWEEN MIDPOINTS OF CYCLE STAGES

IN MONTHS POSITIVE PLAN

CYCLE DATES			I-II	II-III	III-IV	IV-V	V-VI	VI-VII	VII-VIII	VIII-
1933 3	1937 5	1938 6	8.5	16.5	16.5	8.5	2.5	4.0	4.0	[illegible]
1938 6	1945 2	1945 10	13.5	26.5	26.5	13.5	1.5	2.5	2.5	[illegible]
1945 10	1948 11	1949 10	6.5	12.0	12.0	6.5	2.0	3.5	3.5	[illegible]
1949 10	1953 7	1954 8	8.0	14.5	14.5	8.0	2.5	4.0	4.0	
1954 8	1957 7	1958 4	6.0	11.5	11.5	6.0	2.0	2.5	2.5	
1958 4	1960 5	1961 2	4.5	8.0	8.0	4.5	2.0	2.5	2.5	
TOTAL			47.0	89.0	89.0	47.0	12.5	19.0	19.0	[illegible]
AV.			7.8	14.8	14.8	7.8	2.1	3.2	3.2	

Output Table 3B-8

UNEMPLOYED RATE, NICB, CENSUS (INVERTED) 8292

PER CENT X 100

EFERENCE CYCLE ANALYSIS

MEASURES OF SECULAR MOVEMENTS POSITIVE PLAN

CYCLE DATES						AVERAGE MONTHLY STANDING			ABSOLUTE CHANGE FROM PRECEDING PHASE		ABSOLUTE CHANGE FROM PRECEDING CYCLE	
ROUGH		PEAK		TROUGH		EXP-N	CONT-N	CYCLE	EXP-N	CONT-N	TOTAL	PER MONTH
927	11	1929	8	1933	3	.0	1487.2	.0	.0	.0	.0	.00
933	3	1937	5	1938	6	1944.4	1646.2	1882.9	457.2	-298.2	.0	.00
938	6	1945	2	1945	10	969.1	182.4	897.6	-677.2	-786.6	-985.3	-13.05
45	10	1948	11	1949	10	401.9	555.9	437.2	219.4	154.0	-460.4	-6.77
49	10	1953	7	1954	8	387.7	459.0	403.7	-168.2	71.3	-33.5	-.63
54	8	1957	7	1958	4	442.1	535.0	461.1	-16.9	92.9	57.5	1.13
58	4	1960	5	1961	2	595.2	598.4	596.0	60.2	3.3	134.9	3.46
OTAL								4678.4			-1286.8	-15.87
ERAGE								779.7			-257.4	-3.17
ERAGE DEVIATION								407.0			372.4	5.39
GHTED AVERAGE												-4.492

Output Table 3B-9

DURATION OF CYCLICAL MOVEMENTS IN MONTHS

CYCLE DATES						EXPANSION	CONTRACTION	FULL CYCLE
1933	3	1937	5	1938	6	50	13	63
1938	6	1945	2	1945	10	80	8	88
1945	10	1948	11	1949	10	37	11	48
1949	10	1953	7	1954	8	45	13	58
1954	8	1957	7	1958	4	35	9	44
1958	4	1960	5	1961	2	25	9	34
TOTAL						272	63	335
AVERAGE						45.3	10.5	55.8
AVERAGE DEVIATION						13.1	1.8	13.8

Output Table 3B-10

UNEMPLOYED RATE, NICB, CENSUS (INVERTED) 8292

PER CENT X 100 ·

REFERENCE CYCLE ANALYSIS

CYCLICAL AMPLITUDES INVERTED PLAN

CYCLE DATES			CYCLE DEVIATIONS			AMPLITUDES			AMPLITUDES PER MONTH		
PEAK	TROUGH	PEAK	PEAK	TROUGH	PEAK	FALL	RISE	TOTAL	FALL	RISE	TOTAL
1929 8	1933 3	1937 5	-1669.4	1012.0	-465.4	2681.3	-1477.3	4158.7	62.36	-29.55	44.72
1937 5	1938 6	1945 2	203.9	1110.6	-945.7	906.7	-2056.3	2963.0	69.74	-25.70	31.8[illegible]
1945 2	1945 10	1948 11	-244.9	13.1	33.1	258.0	20.0	238.0	32.25	.54	5.2[illegible]
1948 11	1949 10	1953 7	-24.7	278.6	-158.1	303.3	-436.7	740.0	27.58	-9.70	13.2[illegible]
1953 7	1954 8	1957 7	-184.0	146.3	-24.0	330.3	-170.3	500.7	25.41	-4.87	10.4[illegible]
1957 7	1958 4	1960 5	-156.6	136.4	-52.6	293.0	-189.0	482.0	32.55	-7.56	14.1[illegible]
TOTAL			-2075.6	2697.1	-1612.6	4772.7	-4309.7	9082.4	249.89	-76.84	119.6[illegible]
AVERAGE			-345.9	449.5	-268.8	795.4	-718.3	1513.7	41.65	-12.81	19.9[illegible]
AVERAGE DEVIATION			441.1	407.9	291.2	665.7	699.0	1364.7	16.27	9.88	12.2[illegible]
WEIGHTED AVERAGE									49.20	-15.84	24.6[illegible]

Output Table 3B-11

CYCLE PATTERNS INVERTED PLAN
STANDINGS, DEVIATIONS FROM CYCLE BASE

CYCLE DATES			PEAK	II	III	IV	TROUGH	VI	VII	VIII	PEAK	CYCLE
1929 8	1933 3	1937 5	-1669.4				1012.0				-465.4	1733.[illegible]
				-1298.0	-216.7	783.8		542.6	312.0	-231.0		
1937 5	1938 6	1945 2	203.9				1110.6				-945.7	1063.[illegible]
				180.0	549.8	1000.0		738.1	-138.2	-889.6		
1945 2	1945 10	1948 11	-244.9				13.1				33.1	362[illegible]
				-247.9	-214.2	-92.9		67.7	42.3	8.6		
1948 11	1949 10	1953 7	-24.7				278.6				-158.1	420[illegible]
				24.9	134.5	234.9		112.6	-95.2	-129.5		
1953 7	1954 8	1957 7	-184.0				146.3				-24.0	446[illegible]
				-143.0	62.3	124.3		43.4	-23.5	-38.0		
1957 7	1958 4	1960 5	-156.6				136.4				-52.6	57[illegible]
				-128.6	-81.3	50.1		119.8	-36.9	-39.3		
TOTAL			-2075.6	-1612.4	234.4	2100.3	2697.1	1624.2	60.6	-1318.7	-1612.6	
AVERAGE			-345.9	-268.7	39.1	350.1	449.5	270.7	10.1	-219.8	-268.8	
AVERAGE DEVIATION			441.1	343.1	209.8	361.3	407.9	246.4	111.4	227.0	291.2	

Output Table 3B-12

UNEMPLOYED RATE, NICB, CENSUS (INVERTED) 8292

PER CENT X 100

EFERENCE CYCLE ANALYSIS

CYCLE PATTERNS INVERTED PLAN

CYCLE DATES			STAGE TO STAGE CHANGE OF STANDINGS, TOTAL CHANGE							
EAK	TROUGH	PEAK	I-II	II-III	III-IV	IV-V	V-VI	VI-VII	VII-VIII	VIII-IX
929 8	1933 3	1937 5	371.3	1081.3	1000.6	228.1	-469.4	-230.6	-542.9	-234.4
937 5	1938 6	1945 2	-23.9	369.8	450.3	110.6	-372.5	-876.3	-751.4	-56.2
945 2	1945 10	1948 11	-3.0	33.7	121.3	106.0	54.6	-25.4	-33.8	24.6
948 11	1949 10	1953 7	49.7	109.6	100.4	43.7	-166.0	-207.8	-34.3	-28.5
953 7	1954 8	1957 7	41.1	205.3	62.0	22.0	-102.9	-66.8	-14.5	13.9
957 7	1958 4	1960 5	28.0	47.3	131.3	86.3	-16.7	-156.6	-2.4	-13.3
OTAL			463.2	1846.9	1865.9	596.7	-1072.9	-1563.6	-1379.3	-293.9
ERAGE			77.2	307.8	311.0	99.5	-178.8	-260.6	-229.9	-49.0
ERAGE DEVIATION			98.0	278.5	276.3	48.8	161.4	205.2	278.2	64.2

Output Table 3B-13

CYCLE PATTERNS INVERTED PLAN

CYCLE DATES			STAGE TO STAGE CHANGE OF STANDINGS, CHANGE PER MONTH							
AK	TROUGH	PEAK	I-II	II-III	III-IV	IV-V	V-VI	VI-VII	VII-VIII	VIII-IX
29 8	1933 3	1937 5	49.5	77.2	71.5	30.4	-55.2	-14.0	-32.9	-27.6
37 5	1938 6	1945 2	-9.6	92.4	112.6	44.2	-27.6	-33.1	-28.4	-7.1
45 2	1945 10	1948 11	-.0	13.5	48.5	70.7	8.4	-2.7	.0	2.5
48 11	1949 10	1953 7	24.8	31.3	28.7	21.8	-20.8	-14.3	-2.8	.0
53 7	1954 8	1957 7	16.4	51.3	15.5	8.8	-17.1	.0	-1.7	1.3
57 7	1958 4	1960 5	14.0	18.9	52.5	43.2	.0	-19.6	.0	-1.3
JT AL			95.2	284.7	329.3	219.1	-112.3	-83.6	-65.7	-32.2
ERAGE			15.9	47.4	54.9	36.5	-18.7	-13.9	-11.0	-5.4
ERAGE DEVIATION			14.4	26.2	24.8	16.2	15.8	8.4	13.1	8.0
GHTED AVERAGE			25.7	60.5	61.2	33.1	-22.8	-17.6	-15.5	-6.3

Output Table 3B-14

INTERVALS BETWEEN MIDPOINTS OF CYCLE STAGES

IN MONTHS INVERTED PLAN

CYCLE DATES			I-II	II-III	III-IV	IV-V	V-VI	VI-VII	VII-VIII	VIII-IX
9 8	1933 3	1937 5	7.5	14.0	14.0	7.5	8.5	16.5	16.5	8.5
7 5	1938 6	1945 2	2.5	4.0	4.0	2.5	13.5	26.5	26.5	13.5
5 2	1945 10	1948 11	1.5	2.5	2.5	1.5	6.5	12.0	12.0	6.5
3 11	1949 10	1953 7	2.0	3.5	3.5	2.0	8.0	14.5	14.5	8.0
3 7	1954 8	1957 7	2.5	4.0	4.0	2.5	6.0	11.5	11.5	6.0
7	1958 4	1960 5	2.0	2.5	2.5	2.0	4.5	8.0	8.0	4.5
AL			18.0	30.5	30.5	18.0	47.0	89.0	89.0	47.0
			3.0	5.1	5.1	3.0	7.8	14.8	14.8	7.8

Output Table 3B-15

UNEMPLOYED RATE, NICB, CENSUS (INVERTED)

PER CENT X 100

REFERENCE CYCLE ANALYSIS

CYCLE DATES			MEASURES OF SECULAR MOVEMENTS					INVERTED PLAN	
			AVERAGE MONTHLY STANDING			ABSOLUTE CHANGE FROM PRECEDING PHASE		ABSOLUTE CHANGE FROM PRECEDI CYCLE	
PEAK	TROUGH	PEAK	CONT-N	EXP-N	CYCLE	CONT-N	EXP-N	TOTAL	PER
1929 8	1933 3	1937 5	1487.2	1944.4	1733.0	.0	457.2	.0	.0
1937 5	1938 6	1945 2	1646.2	969.1	1063.7	-298.2	-677.2	-669.3	-7.2
1945 2	1945 10	1948 11	182.4	401.9	362.9	-786.6	219.4	-700.9	-10.1
1948 11	1949 10	1953 7	555.9	387.7	420.7	154.0	-168.2	57.9	1.1
1953 7	1954 8	1957 7	459.0	442.1	446.7	71.3	-16.9	26.0	.5
1957 7	1958 4	1960 5	535.0	595.2	579.3	92.9	60.2	132.5	3.2
1960 5	1961 2		598.4	.0	.0	3.3	.0	.0	.0
TOTAL					4606.3			-1153.8	-12.4
AVERAGE					767.7			-230.8	-2.5
AVERAGE DEVIATION					420.4			363.5	4.9
WEIGHTED AVERAGE									-3.7

Output Table 3B-16

DURATION OF CYCLICAL MOVEMENTS IN MONTHS

CYCLE DATES			CONTRACTION	EXPANSION	FULL CYCLE
1929 8	1933 3	1937 5	43	50	93
1937 5	1938 6	1945 2	13	80	93
1945 2	1945 10	1948 11	8	37	45
1948 11	1949 10	1953 7	11	45	56
1953 7	1954 8	1957 7	13	35	48
1957 7	1958 4	1960 5	9	25	34
TOTAL			97	272	369
AVERAGE			16.2	45.3	61.5
AVERAGE DEVIATION			8.9	13.1	21.0

Output Table 3B-17

UNEMPLOYED RATE, NICB, CENSUS (INVERTED) 8292

PER CENT X 100

REFERENCE CYCLE ANALYSIS

CONFORMITY TO BUSINESS CYCLES

AVERAGE CHANGE PER MONTH

ON CYCLE DATES TROUGH	PEAK	TROUGH	TROUGH-PEAK-TROUGH DURING EXPANSION	BASIS DURING CONT.N	CON. MINUS PRECED. EXP.	ON CYCLE DATES PEAK	TROUGH	PEAK	PEAK-TROUGH-PEAK DURING CONT.N	BASIS DURING EXPANSION	CON. MINUS SUCCED. EXP.
						1929 8	1933 3	1937 5	62.36	-29.55	91.90
933 3	1937 5	1938 6	-29.55	69.74	99.29	1937 5	1938 6	1945 2	69.74	-25.70	95.45
938 6	1945 2	1945 10	-25.70	32.25	57.95	1945 2	1945 10	1948 11	32.25	.54	31.71
945 10	1948 11	1949 10	.54	27.58	27.04	1948 11	1949 10	1953 7	27.58	-9.70	37.28
949 10	1953 7	1954 8	-9.70	25.41	35.11	1953 7	1954 8	1957 7	25.41	-4.87	30.28
954 8	1957 7	1958 4	-4.87	32.55	37.42	1957 7	1958 4	1960 5	32.55	-7.56	40.11
958 4	1960 5	1961 2	-7.56	16.96	24.52						
ERAGE			-12.81	34.08	46.89				41.65	-12.81	54.45
ERAGE DEVIATION			9.88	11.89	21.15				16.27	9.88	26.15
TIO OF CONFORMITY TO REFERENCE											
PANSIONS			-.667							-.667	
NTRACTIONS				-1.000					-1.000		
LL CYCLES, TROUGH-TO-TROUGH					-1.000						
LL CYCLES,PEAK-TO-PEAK											-1.000
L CYCLES,BOTH WAYS					-1.000						

FOR B-4 ANALYSIS

FIRST MONTH ESTIMATED AT 129

Output Table 3B-18

MING DIFFS RECOGNIZED BY SHIFTING REF. DATES,TROUGH 2 MONTHS, PEAK -4 MONTHS

CYCLICAL AMPLITUDES POSITIVE PLAN

CYCLE DATES JGH	PEAK	TROUGH	CYCLE DEVIATIONS TROUGH	PEAK	TROUGH	AMPLITUDES RISE	FALL	TOTAL	AMPLITUDES PER MONTH RISE	FALL	TOTAL
3 5	1937 1	1938 8	928.1	-510.3	240.1	-1438.3	750.3	-2188.7	-32.69	39.49	-34.74
3 8	1944 10	1945 12	1244.8	-744.2	-439.6	-1989.0	304.7	-2293.7	-26.88	21.76	-26.06
5 12	1948 7	1949 12	-31.9	-80.2	221.4	-48.3	301.7	-350.0	-1.56	17.75	-7.29
9 12	1953 3	1954 10	271.1	-132.6	173.1	-403.7	305.7	-709.3	-10.35	16.09	-12.23
10	1957 3	1958 6	105.9	-67.4	272.9	-173.3	340.3	-513.7	-5.98	22.69	-11.67
6	1960 1	1961 4	147.2	-77.8	103.8	-225.0	181.7	-406.7	-11.84	12.11	-11.96
AL			2665.2	-1612.5	571.8	-4277.7	2184.3	-6462.0	-89.29	129.88	-103.96
AGE			444.2	-268.8	95.3	-712.9	364.1	-1077.0	-14.88	21.65	-17.33
AGE DEVIATION			428.2	239.0	178.3	667.1	128.8	776.1	9.93	6.33	8.72
HTED AVERAGE									-18.13	22.06	-19.29

Output Table 3B-19

UNEMPLOYED RATE, NICB, CENSUS (INVERTED) 829.

PER CENT X 100

REFERENCE CYCLE ANALYSIS

TIMING DIFFS RECOGNIZED BY SHIFTING REF. DATES,TROUGH 2 MONTHS, PEAK -4 MONTHS

DURATION OF CYCLICAL MOVEMENTS IN MONTHS

CYCLE DATES						EXPANSION	CONTRACTION	FULL CYCLE
1933	5	1937	1	1938	8	44	19	63
1938	8	1944	10	1945	12	74	14	88
1945	12	1948	7	1949	12	31	17	48
1949	12	1953	3	1954	10	39	19	58
1954	10	1957	3	1958	6	29	15	44
1958	6	1960	1	1961	4	19	15	34
TOTAL						236	99	335
AVERAGE						39.3	16.5	55.8
AVERAGE DEVIATION						13.1	1.8	13.8

Output Table 3B-20

CYCLICAL AMPLITUDES INVERTED PLAN

CYCLE DATES						CYCLE DEVIATIONS			AMPLITUDES			AMPLITUDES PER MONT[H]		
PEAK		TROUGH		PEAK		PEAK	TROUGH	PEAK	FALL	RISE	TOTAL	FALL	RISE	TOTAL
1929	4	1933	5	1937	1	-1554.2	1107.4	-330.9	2661.7	-1438.3	4100.0	54.32	-32.69	44.
1937	1	1938	8	1944	10	237.2	987.5	-1001.5	750.3	-1989.0	2739.3	39.49	-26.88	29.
1944	10	1945	12	1948	7	-225.9	78.7	30.4	304.7	-48.3	353.0	21.76	-1.56	7.
1948	7	1949	12	1953	3	-59.0	242.7	-161.0	301.7	-403.7	705.3	17.75	-10.35	12
1953	3	1954	10	1957	3	-167.1	138.5	-34.8	305.7	-173.3	479.0	16.09	-5.98	9
1957	3	1958	6	1960	1	-166.3	174.0	-51.0	340.3	-225.0	565.3	22.69	-11.84	16
TOTAL						-1935.4	2729.0	-1548.7	4664.4	-4277.7	8942.0	172.09	-89.29	120
AVERAGE						-322.6	454.8	-258.1	777.4	-712.9	1490.3	28.68	-14.88	20
AVERAGE DEVIATION						410.6	395.1	272.0	628.1	667.1	1286.2	12.15	9.93	11
WEIGHTED AVERAGE												35.07	-18.13	24

Output Table 3B-21

UNEMPLOYED RATE, NICB, CENSUS (INVERTED) 8292

PER CENT X 100

EFERENCE CYCLE ANALYSIS

TIMING DIFFS RECOGNIZED BY SHIFTING REF. DATES,TROUGH 2 MONTHS, PEAK -4 MONTHS

CONFORMITY TO BUSINESS CYCLES

AVERAGE CHANGE PER MONTH

	CYCLE ON DATES			TROUGH-PEAK-TROUGH	BASIS		CYCLE ON DATES			PEAK-TROUGH-PEAK	BASIS	
	ROUGH	PEAK	TROUGH	DURING EXPANSION	DURING CONT.N	CON. MINUS PRECED. EXP.	PEAK	TROUGH	PEAK	DURING CONT.N	DURING EXPANSION	CON. MINUS SUCCED. EXP.
							1929 4	1933 5	1937 1	54.32	-32.69	87.01
	'33 5	1937 1	1938 8	-32.69	39.49	72.18	1937 1	1938 8	1944 10	39.49	-26.88	66.37
	'38 8	1944 10	1945 12	-26.88	21.76	48.64	1944 10	1945 12	1948 7	21.76	-1.56	23.32
	45 12	1948 7	1949 12	-1.56	17.75	19.30	1948 7	1949 12	1953 3	17.75	-10.35	28.09
	49 12	1953 3	1954 10	-10.35	16.09	26.44	1953 3	1954 10	1957 3	16.09	-5.98	22.06
	54 10	1957 3	1958 6	-5.98	22.69	28.67	1957 3	1958 6	1960 1	22.69	-11.84	34.53
	58 6	1960 1	1961 4	-11.84	12.11	23.95						
ERAGE				-14.88	21.65	36.53				28.68	-14.88	43.56
ERAGE DEVIATION				9.93	6.33	15.92				12.15	9.93	22.08
IO OF CONFORMITY TO REFERENCE												
ANSIONS				-1.000							-1.000	
TRACTIONS					-1.000					-1.000		
L CYCLES, TROUGH-TO-TROUGH						-1.000						
L CYCLES,PEAK-TO-PEAK												-1.000
L CYCLES,BOTH WAYS							-1.000					

Output Table 3B-22

NEMPLOYED RATE, NICB, CENSUS (INVERTED) 8292

ER CENT X 100

CIFIC CYCLE ANALYSIS

CYCLICAL AMPLITUDES POSITIVE PLAN

	CYCLE DATES			CYCLE DEVIATIONS			AMPLITUDES			AMPLITUDES PER MONTH		
	GH	PEAK	TROUGH	TROUGH	PEAK	TROUGH	RISE	FALL	TOTAL	RISE	FALL	TOTAL
	3 5	1937 7	1938 6	937.6	-620.7	321.6	-1558.3	942.3	-2500.7	-31.17	85.67	-40.99
	6	1944 10	1946 5	1311.4	-749.6	-416.6	-2061.0	333.0	-2394.0	-27.12	17.53	-25.20
	5	1948 1	1949 10	7.7	-69.0	260.7	-76.7	329.7	-406.3	-3.83	15.70	-9.91
	10	1953 5	1954 9	292.2	-140.8	188.2	-433.0	329.0	-762.0	-10.07	20.56	-12.92
	9	1957 3	1958 7	119.3	-76.1	264.6	-195.3	340.7	-536.0	-6.51	21.29	-11.65
	7	1959 6	1961 5	148.6	-82.4	106.6	-231.0	189.0	-420.0	-21.00	8.22	-12.35
AL				2816.8	-1738.6	725.1	-4555.3	2463.7	-7019.0	-99.70	168.96	-113.02
AGE				469.5	-289.8	120.9	-759.2	410.6	-1169.8	-16.62	28.16	-18.84
AGE DEVIATION				436.7	263.6	183.9	700.3	177.2	851.7	9.81	19.17	9.51
TED AVERAGE										-19.81	23.24	-20.89

Output Table 3B-23

82

UNEMPLOYED RATE, NICB, CENSUS (INVERTED)

PER CENT X 100

SPECIFIC CYCLE ANALYSIS

CYCLE PATTERNS POSITIVE PLAN

STANDINGS, IN ORIGINAL UNITS

CYCLE DATES												
TROUGH	PEAK	TROUGH	TROUGH	II	III	IV	PEAK	VI	VII	VIII	TROUGH	CYCLE
1933 5	1937 7	1938 6	2790.3	2211.7	1969.2	1451.7	1232.0	1281.7	1769.8	2096.3	2174.3	1852
1938 6	1944 10	1946 5	2174.3	1808.2	1023.0	206.0	113.3	114.2	225.3	430.2	446.3	863
1946 5	1948 1	1949 10	446.3	425.5	402.3	389.0	369.7	376.0	413.8	598.3	699.3	438
1949 10	1953 5	1954 9	699.3	544.7	330.9	295.9	266.3	277.2	478.0	577.0	595.3	407
1954 9	1957 3	1958 7	595.3	477.9	417.9	419.3	400.0	415.0	499.4	704.2	740.7	476
1958 7	1959 6	1961 5	740.7	710.3	607.8	533.0	509.7	544.6	536.6	661.3	698.7	592

Output Table 3B-24

CYCLE PATTERNS POSITIVE PLAN
STANDINGS, DEVIATIONS FROM CYCLE BASE

CYCLE DATES												
TROUGH	PEAK	TROUGH	TROUGH	II	III	IV	PEAK	VI	VII	VIII	TROUGH	CYCLE
1933 5	1937 7	1938 6	937.6				-620.7				321.6	185
				359.0	116.5	-401.0		-571.1	-83.0	243.6		
1938 6	1944 10	1946 5	1311.4				-749.6				-416.6	86
				945.2	160.1	-656.9		-748.8	-637.6	-432.8		
1946 5	1948 1	1949 10	7.7				-69.0				260.7	43
				-13.2	-36.4	-49.7		-62.7	-24.8	159.6		
1949 10	1953 5	1954 9	292.2				-140.8				188.2	4
				137.6	-76.2	-111.2		-129.9	70.9	169.9		
1954 9	1957 3	1958 7	119.3				-76.1				264.6	4
				1.8	-58.2	-56.8		-61.1	23.3	228.1		
1958 7	1959 6	1961 5	148.6				-82.4				106.6	5
				118.3	15.7	-59.0		-47.5	-55.4	69.3		
TOTAL			2816.8	1548.8	121.6	-1334.6	-1738.6	-1621.0	-706.6	437.7	725.1	
AVERAGE			469.5	258.1	20.3	-222.4	-289.8	-270.2	-117.8	72.9	120.9	
AVERAGE DEVIATION			436.7	262.6	78.7	204.4	263.6	259.9	173.3	169.8	183.9	

Output Table 3B-25

:NEMPLOYED RATE, NICB, CENSUS (INVERTED) 8292

'ER CENT X 100

CIFIC CYCLE ANALYSIS

CYCLE PATTERNS POSITIVE PLAN

CYCLE DATES			STAGE TO STAGE CHANGE OF STANDINGS, TOTAL CHANGE							
UGH	PEAK	TROUGH	I-II	II-III	III-IV	IV-V	V-VI	VI-VII	VII-VIII	VIII-IX
3 5	1937 7	1938 6	-578.6	-242.5	-517.5	-219.7	49.7	488.1	326.6	78.0
8 6	1944 10	1946 5	-366.1	-785.2	-817.0	-92.7	.8	111.2	204.8	16.2
6 5	1948 1	1949 10	-20.8	-23.2	-13.3	-19.3	6.3	37.8	184.5	101.0
10	1953 5	1954 9	-154.6	-213.8	-35.0	-29.6	10.9	200.8	99.0	18.3
9	1957 3	1958 7	-117.4	-60.0	1.4	-19.3	15.0	84.4	204.8	36.5
7	1959 6	1961 5	-30.3	-102.6	-74.8	-23.3	34.9	-7.9	124.7	37.4
AL			-1268.0	-1427.2	-1456.2	-404.0	117.6	914.3	1144.3	287.4
AGE			-211.3	-237.9	-242.7	-67.3	19.6	152.4	190.7	47.9
AGE DEVIATION			174.0	184.0	283.0	59.2	15.1	128.0	54.7	27.8

Output Table 3B-26

CYCLE PATTERNS POSITIVE PLAN

CYCLE DATES			STAGE TO STAGE CHANGE OF STANDINGS, CHANGE PER MONTH							
GH	PEAK	TROUGH	I-II	II-III	III-IV	IV-V	V-VI	VI-VII	VII-VIII	VIII-IX
5	1937 7	1938 6	-68.1	-14.7	-31.4	-25.8	24.8	139.5	93.3	39.0
6	1944 10	1946 5	-28.2	-31.4	-32.7	-4.3	.6	18.5	34.1	6.9
5	1948 1	1949 10	.0	-1.2	-3.1	-2.3	.0	1.2	28.4	25.3
10	1953 5	1954 9	-20.6	-15.3	.0	-1.1	2.7	40.2	19.8	5.3
9	1957 3	1958 7	-21.4	-2.9	2.1	-3.3	.0	16.9	41.0	12.1
7	1959 6	1961 5	-15.2	-29.3	-21.4	-11.7	8.7	-1.6	16.6	9.3
AL			-153.4	-94.8	-86.4	-48.4	36.8	214.6	233.2	97.9
GE			-25.6	-15.8	-14.4	-8.1	6.1	35.8	38.9	16.3
GE DEVIATION			15.0	9.7	14.1	7.1	7.1	36.0	18.8	10.5
TED AVERAGE			-31.7	-19.0	-19.4	-10.1	6.0	27.3	34.2	14.7

Output Table 3B-27

INTERVALS BETWEEN MIDPOINTS OF CYCLE STAGES

IN MONTHS POSITIVE PLAN

CYCLE DATES			I-II	II-III	III-IV	IV-V	V-VI	VI-VII	VII-VIII	VIII-IX
5	1937 7	1938 6	8.5	16.5	16.5	8.5	2.0	3.5	3.5	2.0
6	1944 10	1946 5	13.0	25.0	25.0	13.0	3.5	6.0	6.0	3.5
5	1948 1	1949 10	3.5	6.5	6.5	3.5	4.0	6.5	6.5	4.0
10	1953 5	1954 9	7.5	14.0	14.0	7.5	3.0	5.0	5.0	3.0
9	1957 3	1958 7	5.5	9.5	9.5	5.5	3.0	5.0	5.0	3.0
7	1959 6	1961 5	2.0	3.5	3.5	2.0	4.0	7.5	7.5	4.0
-			40.0	75.0	75.0	40.0	19.5	33.5	33.5	19.5
			6.7	12.5	12.5	6.7	3.3	5.6	5.6	3.3

Output Table 3B-28

UNEMPLOYED RATE, NICB, CENSUS (INVERTED)

PER CENT X 100

SPECIFIC CYCLE ANALYSIS

MEASURES OF SECULAR MOVEMENTS POSITIVE PLAN

CYCLE DATES			AVERAGE MONTHLY STANDING			ABSOLUTE CHANGE FROM PRECEDING PHASE		ABSOLUTE CHANGE FROM PRECED[illegible] CYCLE	
TROUGH	PEAK	TROUGH	EXP-N	CONT-N	CYCLE	EXP-N	CONT-N	TOTAL	P[illegible]
1933 5	1937 7	1938 6	1882.0	1719.6	1852.7	.0	-162.4	.0	
1938 6	1944 10	1946 5	1014.2	258.0	863.0	-705.4	-756.2	-989.8	-12[illegible]
1946 5	1948 1	1949 10	405.8	470.0	438.7	147.8	64.3	-424.3	-6[illegible]
1949 10	1953 5	1954 9	393.5	443.6	407.1	-76.5	50.1	-31.6	-[illegible]
1954 9	1957 3	1958 7	441.2	541.5	476.1	-2.4	100.4	69.0	1[illegible]
1958 7	1959 6	1961 5	617.1	580.0	592.0	75.6	-37.1	116.0	2[illegible]
TOTAL					4629.6			-1260.7	-1[illegible]
AVERAGE					771.6			-252.1	-3[illegible]
AVERAGE DEVIATION					390.8			363.9	[illegible]
WEIGHTED AVERAGE									-4.[illegible]

Output Table 3B-29

DURATION OF CYCLICAL MOVEMENTS IN MONTHS

CYCLE DATES			EXPANSION	CONTRACTION	FULL CYCLE
1933 5	1937 7	1938 6	50	11	61
1938 6	1944 10	1946 5	76	19	95
1946 5	1948 1	1949 10	20	21	41
1949 10	1953 5	1954 9	43	16	59
1954 9	1957 3	1958 7	30	16	46
1958 7	1959 6	1961 5	11	23	34
TOTAL			230	106	336
AVERAGE			38.3	17.7	56.0
AVERAGE DEVIATION			18.0	3.3	15.7

Output Table 3B-30

CYCLICAL AMPLITUDES INVERTED PLAN

CYCLE DATES			CYCLE DEVIATIONS			AMPLITUDES			AMPLITUDES	
PEAK	TROUGH	PEAK	PEAK	TROUGH	PEAK	FALL	RISE	TOTAL	FALL	RISE
1937 7	1938 6	1944 10	128.6	1070.9	-990.1	942.3	-2061.0	3003.3	85.67	-27.1[illegible]
1944 10	1946 5	1948 1	-220.4	112.6	35.9	333.0	-76.7	409.7	17.53	-3.8[illegible]
1948 1	1949 10	1953 5	-49.0	280.7	-152.3	329.7	-433.0	762.7	15.70	-10.0[illegible]
1953 5	1954 9	1957 3	-175.7	153.3	-42.0	329.0	-195.3	524.3	20.56	-6.5[illegible]
1957 3	1958 7	1959 6	-172.3	168.3	-62.7	340.7	-231.0	571.7	21.29	-21.0[illegible]
TOTAL			-488.8	1785.9	-1211.1	2274.7	-2997.0	5271.7	160.74	-68.5[illegible]
AVERAGE			-97.8	357.2	-242.2	454.9	-599.4	1054.3	32.15	-13.7[illegible]
AVERAGE DEVIATION			110.1	285.5	299.1	195.0	584.6	779.6	21.41	8.2[illegible]
WEIGHTED AVERAGE									27.41	-16.[illegible]

Output Table 3B-31

UNEMPLOYED RATE, NICB, CENSUS (INVERTED) 8292

PER CENT X 100

ECIFIC CYCLE ANALYSIS

CYCLE PATTERNS INVERTED PLAN
STANDINGS, DEVIATIONS FROM CYCLE BASE

CYCLE DATES	PEAK	II	III	IV	TROUGH	VI	VII	VIII	PEAK	CYCLE
37 7 1938 6 1944 10	128.6				1070.9				-990.1	1103.4
		178.3	666.3	992.9		704.8	-80.4	-897.4		
44 10 1946 5 1948 1	-220.4				112.6				35.9	333.8
		-219.6	-108.4	96.4		91.7	68.5	55.2		
48 1 1949 10 1953 5	-49.0				280.7				-152.3	418.6
		-42.6	-4.8	179.7		126.1	-87.7	-122.7		
53 5 1954 9 1957 3	-175.7				153.3				-42.0	442.0
		-164.8	36.0	135.0		35.9	-24.1	-22.7		
57 3 1958 7 1959 6	-172.3				168.3				-62.7	572.3
		-157.3	-72.9	131.9		138.0	35.4	-39.3		
TAL	-488.8	-406.1	516.2	1535.9	1785.9	1096.5	-88.2	-1026.9	-1211.1	
RAGE	-97.8	-81.2	103.2	307.2	357.2	219.3	-17.6	-205.4	-242.2	
RAGE DEVIATION	110.1	119.2	225.2	274.3	285.5	194.2	55.7	276.8	299.1	

Output Table 3B-32

CYCLE PATTERNS INVERTED PLAN

STAGE TO STAGE CHANGE OF STANDINGS, TOTAL CHANGE

CYCLE DATES TROUGH PEAK	I-II	II-III	III-IV	IV-V	V-VI	VI-VII	VII-VIII	VIII-IX
7 1938 6 1944 10	49.7	488.1	326.6	78.0	-366.1	-785.2	-817.0	-92.7
10 1946 5 1948 1	.8	111.2	204.8	16.2	-20.8	-23.2	-13.3	-19.3
1 1949 10 1953 5	6.3	37.8	184.5	101.0	-154.6	-213.8	-35.0	-29.6
5 1954 9 1957 3	10.9	200.8	99.0	18.3	-117.4	-60.0	1.4	-19.3
3 1958 7 1959 6	15.0	84.4	204.8	36.5	-30.3	-102.6	-74.8	-23.3
AL	82.7	922.3	1019.7	250.0	-689.4	-1184.8	-938.6	-184.3
AGE	16.5	184.5	203.9	50.0	-137.9	-236.9	-187.7	-36.9
AGE DEVIATION	13.3	128.0	49.8	31.6	98.0	219.3	251.7	22.3

Output Table 3B-33

CYCLE PATTERNS INVERTED PLAN

STAGE TO STAGE CHANGE OF STANDINGS, CHANGE PER MONTH

CYCLE DATES TROUGH PEAK	I-II	II-III	III-IV	IV-V	V-VI	VI-VII	VII-VIII	VIII-IX
7 1938 6 1944 10	24.8	139.5	93.3	39.0	-28.2	-31.4	-32.7	-4.3
10 1946 5 1948 1	.6	18.5	34.1	6.9	.0	-1.2	-3.1	-2.3
1 1949 10 1953 5	.0	1.2	28.4	25.3	-20.6	-15.3	.0	-1.1
5 1954 9 1957 3	2.7	40.2	19.8	5.3	-21.4	-2.9	2.1	-3.3
3 1958 7 1959 6	.0	16.9	41.0	12.1	-15.2	-29.3	-21.4	-11.7
L	28.1	216.2	216.6	88.6	-85.3	-80.1	-55.0	-22.6
GE	5.6	43.2	43.3	17.7	-17.1	-16.0	-11.0	-4.5
GE DEVIATION	7.7	38.5	20.0	11.5	7.6	11.5	12.8	2.9
TED AVERAGE	5.3	35.5	39.2	16.1	-21.9	-20.3	-16.0	-5.8

Output Table 3B-34

UNEMPLOYED RATE, NICB, CENSUS (INVERTED)

PER CENT X 100

SPECIFIC CYCLE ANALYSIS

CYCLE DATES	I-II	II-III	III-IV	IV-V	V-VI	VI-VII	VII-VIII	VIII-
	INTERVALS BETWEEN MIDPOINTS OF CYCLE STAGES IN MONTHS INVERTED PLAN							
1937 7 1938 6 1944 10	2.0	3.5	3.5	2.0	13.0	25.0	25.0	1
1944 10 1946 5 1948 1	3.5	6.0	6.0	3.5	3.5	6.5	6.5	
1948 1 1949 10 1953 5	4.0	6.5	6.5	4.0	7.5	14.0	14.0	
1953 5 1954 9 1957 3	3.0	5.0	5.0	3.0	5.5	9.5	9.5	
1957 3 1958 7 1959 6	3.0	5.0	5.0	3.0	2.0	3.5	3.5	
TOTAL	15.5	26.0	26.0	15.5	31.5	58.5	58.5	3
AV.	3.1	5.2	5.2	3.1	6.3	11.7	11.7	

Output Table 3B-35

MEASURES OF SECULAR MOVEMENTS INVERTED PLAN

CYCLE DATES PEAK	TROUGH	PEAK	AVERAGE MONTHLY STANDING CONT-N	EXP-N	CYCLE	ABSOLUTE CHANGE FROM PRECEDING PHASE CONT-N	EXP-N	ABSOLUTE CHANGE FROM PREC CYCLE TOTAL	
1937 7	1938 6	1944 10	1719.6	1014.2	1103.4	-162.4	-705.4	.0	
1944 10	1946 5	1948 1	258.0	405.8	333.8	-756.2	147.8	-769.6	-
1948 1	1949 10	1953 5	470.0	393.5	418.6	64.3	-76.5	84.9	
1953 5	1954 9	1957 3	443.6	441.2	442.0	50.1	-2.4	23.4	
1957 3	1958 7	1959 6	541.5	617.1	572.3	100.4	75.6	130.3	
1959 6	1961 5		580.0	.0	.0	-37.1	.0	.0	
TOTAL					2870.1			-531.1	
AVERAGE					574.0			-132.8	
AVERAGE DEVIATION					211.8			318.4	
WEIGHTED AVERAGE									-

Output Table 3B-36

DURATION OF CYCLICAL MOVEMENTS IN MONTHS

CYCLE DATES	CONTRACTION	EXPANSION	FULL CYCLE
1937 7 1938 6 1944 10	11	76	87
1944 10 1946 5 1948 1	19	20	39
1948 1 1949 10 1953 5	21	43	64
1953 5 1954 9 1957 3	16	30	46
1957 3 1958 7 1959 6	16	11	27
TOTAL	83	180	263
AVERAGE	16.6	36.0	52.6
AVERAGE DEVIATION	2.7	18.8	18.3

4

RECESSION AND RECOVERY ANALYSIS

RATIONALE AND APPROACH

PURPOSE AND USES

RECESSION ANALYSIS AND RECOVERY ANALYSIS are two analogous approaches to the understanding of business cycles.[1] They are primarily designed to facilitate the evaluation of prevailing business conditions by comparing current contractions, or current expansions, with corresponding phases in the past. This is done by measuring changes of individual time series from their standing at cyclical turns and comparing current with past changes over a series of widening time spans. All comparisons are based on seasonally adjusted data, if such adjustment is warranted.

Some illustrations will clarify the simple procedures. Table 11 contains percentage changes of nonagricultural employment[2] from business cycle peaks (three-month average, centered at the peak). For each contraction since 1929, changes are shown over successive spans, varying from six months before[3] to thirty months after a business

[1] The basic approach has been developed by Geoffrey H. Moore. See his *Measuring Recessions,* New York, NBER, 1958. This and other books and papers cited in this chapter are recommended to all users who wish to acquire thorough familiarity with the analysis.

[2] As in the earlier parts of this study, the term nonagricultural employment is used as a short designation for "number of employees in nonagricultural establishments," which is the full title of the series collected and published by the Bureau of Labor Statistics, U.S. Department of Labor.

[3] For the spans *before* the peak, the term "percentage change *from* peak" implies the wrong direction. Percentage deviation from peak levels (or from trough levels) would avoid the directional connotation. However, in this study we shall conform to the terminology used in the basic publications.

TABLE

RECESSION ANALYSIS, NONAGRICULTURAL BUSINESS CYCLE

Date of Peak[a]	*Percentage Change (months before peak)*		*Standing at Peak*[b] *(thousands)*	*Percentage Change*			
	6	*3*		*3*	*6*	*9*	*12*
Aug. 1929	−1.9	−0.8	33,222	−1.5	−4.6	−6.5	−9.7
May 1937	−3.2	−1.3	31,904	+0.6	−2.2	−7.1	−9.7
Feb. 1945	−0.1	−0.4	41,740	−1.5	−3.4	−7.0	−6.0
Nov. 1948	−0.9	−0.1	45,077	−1.4	−2.7	−3.5	−4.2
July 1957	−0.3	+0.1	53,011	−0.6	−1.8	−3.9	−3.9
July 1953	−0.6	+0.1	50,378	−0.5	−2.0	−2.7	−3.3
May 1960	−1.6	0	54,407	−0.4	−1.0	−1.9	−1.3
Average	−1.2	−0.3		−0.7	−2.5	−4.7	−5.5
Avg. deviation	0.9	0.4		0.6	0.9	1.9	2.6
Correlation coefficients (Pearsonian) partial vs. total amplitudes				+0.33	+0.85	+0.61	+0.78

Source: Output Tables 3A-3 and 3A-6.

[a] In order of contraction amplitudes (see next to last column).

cycle peak. Only spans in multiples of three are shown. Chart 14 is an equivalent graph of the percentage changes, except that all monthly spans from −6 months to +22 months are charted and the recessions beginning in 1945 and 1948 are omitted.[4] These presentations permit a comparative evaluation of a current cyclical decline in employment against the background of past employment changes during comparable recession periods. As a recession proceeds, the characteristics of a given activity will emerge with increasing clarity. Similar comparisons can, of course, also be made for expansions. In the simplest though not necessarily the most instructive form of recovery analysis, percentage increases are computed from past business cycle troughs

[4] In order to avoid crowding the chart, we omitted the two war-affected recessions. They seemed to be least relevant for the evaluation of recent and prospective contractions.

11

EMPLOYMENT, PERCENTAGE CHANGE FROM PEAKS, 1929–63

(months after peak)						Percentage Change to Business Cycle Trough[b]	Duration of Recession (months)
15	18	21	24	27	30		
−12.3	−14.3	−15.7	−18.2	−21.3	−23.5	−30.7	43
−9.3	−7.1	−6.2	−5.4	−3.8	−1.2	−10.0	13
−1.1	+1.8	+3.6	+4.4	+4.4	+5.1	−7.6	8
−4.2	−1.0	+2.2	+3.5	+5.4	+6.1	−4.1	11
−3.0	−1.1	+0.5	+1.3	+0.4	+2.2	−3.8	9
−3.0	−2.0	−0.5	+1.0	+1.9	+3.0	−3.3	13
−0.3	+0.4	+1.1	+2.0	+2.4	+2.7	−1.8	9
−4.8	−3.3	−2.2	−1.6	−1.5	−0.8	−8.8	15.1
3.5	4.2	5.0	5.8	6.3	6.6	6.6	7.9
+0.85	+0.91	+0.92	+0.94	+0.95	+0.96		−0.96[c]

[b] Three-month average centered at turn.

[c] Correlation of duration with recession amplitude.

over successively increasing spans. Comparisons of a current expansion may then be made with preceding expansions, numerically and graphically, as is shown for nonagricultural employment in Table 12 and in Chart 15. The characteristics of a current cyclical upswing in employment, particularly its relative briskness, will become increasingly apparent as the expansion proceeds. A set of such comparisons for a variety of strategic economic activities enables gauging, and perhaps anticipating, the pace of a general economic recovery (or recession) while it is in process.

Recession-recovery analysis, in common with the business cycle analysis described earlier, elucidates the process of cyclical fluctuations in economic activity by a systematic description of the cyclical behavior of many individual time series. In both approaches the descriptive measures are constructed in such a way that economic be-

CHART 14
RECESSION ANALYSIS,
NONAGRICULTURAL EMPLOYMENT,
PERCENTAGE CHANGE FROM BUSINESS CYCLE PEAKS,
1929–62

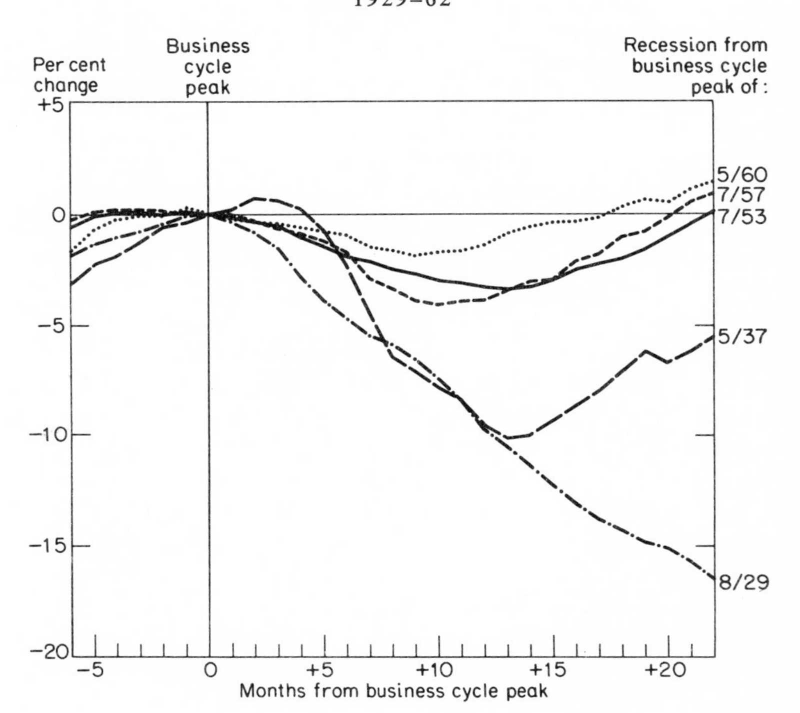

havior can be observed in each historical business cycle, comparisons can be made among activities and among cycles, and generalizations about cyclical behavior in the economy as a whole can be obtained by summing, averaging, and comparing basic measures.

However, recession-recovery analysis differs from the Bureau's business cycle analysis in the goals set, the assumptions stipulated, and the measures derived. The standard business cycle analysis is a historically oriented research tool, largely designed to bring out the characteristics of the cyclical behavior of diverse economic activities, in the expectation that this will provide insights and generalizations about the cyclical process as a whole. Recession-recovery analysis, by contrast, aims mainly at the understanding of current business conditions. It is forecasting-and-policy-oriented, and focuses on the identification of the characteristics of a current process as compared to previous cyclical experience.[5] This difference in orientation is reflected in the choice of the units of observation and measurement. In the standard business cycle analysis, the cycle or cycle phase is the unit of observation, and the data are expressed in terms of cycle averages; standings are computed for cycle stages; time is measured and behavior described in terms of fractions of the length of a completed cyclical phase; comparisons are made and summary measures are derived for the same fractions. These measures are well suited to a research approach that puts no particular premium on currency of results, conventionality of measurement, or accessibility of the analysis to nonspecialists. For the purpose of deriving broad generalizations on cyclical behavior, it is no serious loss if one has to wait until a cycle is completed before it can be included in the cyclical averages. In recession-recovery analysis, by contrast, the basic measures are conventional percentage changes, based on original units; they are computed for months and quarters in chronological sequence. The goal, i.e., provision of up-to-date guidance for the evaluation of current business conditions, is reflected in the use of measures that can be computed before a current cycle phase reaches an end. In fact, recession and recovery analyses are specifically designed to evaluate behavior during current incomplete phases. The only prerequisite is the establishment of past cyclical turns.

RECESSION PATTERNS

What insights can be derived from the comparison of a recent recession with prior ones? Table 11 and Chart 14 show the decline in employment during 1960–61 to be mild compared to the declines during other contractions. This mildness becomes apparent as early as four to six months after the business cycle peak of 1960. It is important

[5] The distinction is perhaps too sharply drawn, since business cycle analysis can also be focused on the distinctive characteristics of a particular cycle, and recession-recovery analysis can also be used to emphasize characteristics of an historical cycle or those common to many cycles. However, the uses described in the text are the prevailing ones, which may explain why recession-recovery analysis was developed during a later, more policy-oriented historical period and why, until now, it did not include computation of changes averaged over all corresponding phases.

TABLE

RECOVERY ANALYSIS, NONAGRICULTURAL BUSINESS CYCLE

Date of Trough[a]	*Percentage Change (months before trough)*		*Standing at Trough*[b] *(thousands)*	*Percentage Change*			
	6	*3*		*3*	*6*	*9*	*12*
June 1938	+6.2	+2.3	28,725	+1.5	+4.2	+5.0	+5.9
Mar. 1933	+2.2	+2.2	23,030	+3.8	+11.3	+12.1	+15.8
Feb. 1961	+1.4	+0.7	53,451	+0.4	+1.4	+2.1	+2.9
Oct. 1945	+7.2	+5.4	38,559	+3.0	+5.8	+8.8	+11.5
Oct. 1949	+2.0	+0.6	43,215	+0.6	+2.5	+5.0	+7.7
Aug. 1954	+1.2	+0.3	48,720	+0.8	+1.7	+3.5	+4.6
Apr. 1958	+3.4	+2.1	50,978	0	+0.9	+2.9	+4.5
Average	+3.4	+1.9		+1.4	+4.0	+5.6	+7.5
Avg. deviation	1.9	1.2		1.1	2.7	2.8	3.5
Correlation coefficients (Pearsonian) partial vs. total amplitudes				+0.57	+0.66	+0.52	+0.46

Source: Output Tables 3A-3 and 3A-6.
[a] In order of expansion amplitudes (see next to last column).
[b] Three-month average centered at turn.

that an early manifestation of the relative steepness of a decline is not confined to the most recent contraction. The depths of the contractions of 1929–32 and 1937–38 can be inferred from the low level of the relatives after about four and seven months, respectively, and one can indeed discern a general association between initial and eventual amplitudes—albeit an association that emerges only gradually, that is imperfect, and that is somewhat obscured by irregular movements. Still it exists and can be utilized in conjunction with other approaches to evaluate cyclical prospects.

If a large number of recessions are compared on one chart, the multitude of curves may be confusing, particularly if the activities show a great deal of irregular movement. A device for depicting the relative impact of a current contraction, without presenting the de-

12

EMPLOYMENT, PERCENTAGE CHANGE FROM TROUGHS, 1932–66

(months after trough)						Percentage Change to Business Cycle Peak[b]	Duration of Expansion (months)
15	18	21	24	27	30		
+8.4	+9.9	+10.0	+11.1	+14.5	+18.8	+45.3	80
+17.4	+15.0	+17.3	+19.7	+20.1	+22.2	+38.5	50
+3.8	+4.2	+4.5	+4.8	+5.7	+6.1	+17.2[c]	61[c]
+12.8	+12.8	+13.2	+14.8	+15.8	+15.0	+16.9	37
+9.4	+10.8	+11.0	+10.6	+11.7	+12.2	+16.6	45
+5.7	+7.0	+7.6	+7.7	+8.3	+8.9	+8.8	35
+5.4	+4.4	+6.5	+7.0	+6.3	+5.9	+6.7	25
+9.0	+9.2	+10.0	+10.8	+11.8	+12.7	21.43	47.6
3.6	3.4	3.3	3.8	4.3	5.1	11.01	13.8
+0.54	+0.58	+0.55	+0.58	+0.72	+0.86		+0.81[d]

[c] To March 1966, last month included.
[d] Correlation of duration with expansion amplitude.

tailed movements of all previous ones, is shown in Chart 16. The positions of the dots show the relative declines of the same activity in previous contractions; the contractions are numbered on the basis of their eventual severity. The solid line shows the behavior of the 1960–61 decline. For further simplification, relatives are shown for every third month only. The initial comparative mildness of the 1960–61 contraction in employment and the gradual confirmation of this mildness until the end of the decline are readily apparent in this presentation.

Similar comparisons can be carried through for a variety of strategic economic activities. Since the severity of a given contraction tends to be reflected in many activities, it is usually possible to classify a current contraction in business conditions as mild, intermediate, or

CHART 15

RECOVERY ANALYSIS,
NONAGRICULTURAL EMPLOYMENT,
PERCENTAGE CHANGE FROM BUSINESS CYCLE TROUGHS,
1932–63

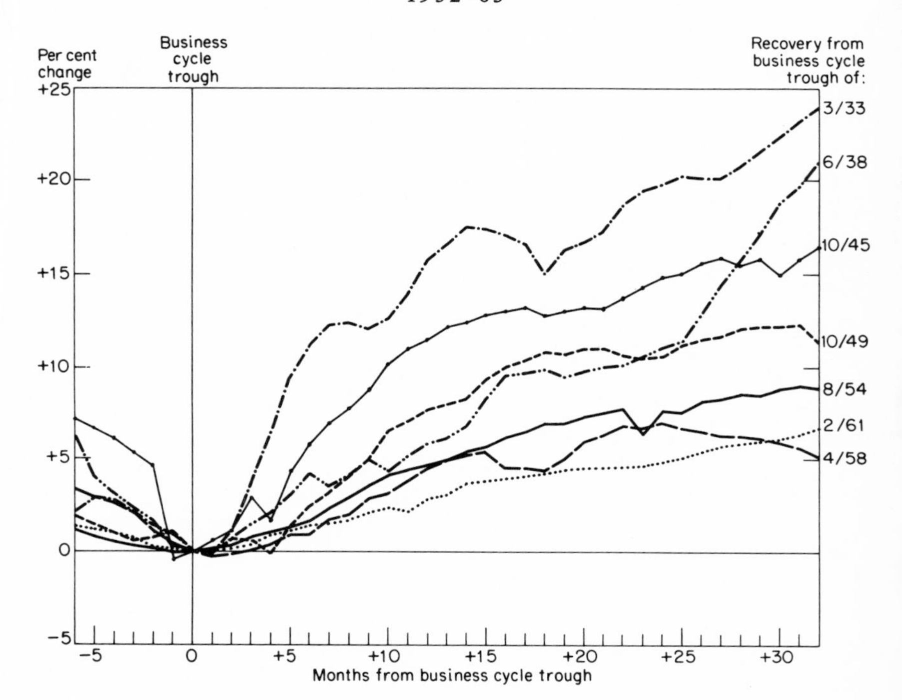

severe—at least for the span over which current observations are available. Historically, initial and full amplitudes tend to be correlated. To the degree that this association is maintained, the rough classification may hold for the recession as a whole.

The relationship between full and partial amplitude for various recessions can be described by means of correlation coefficients. Table 11 presents the full decline of nonagricultural employment during each business cycle contraction from 1929 on, in order of severity. The change in employment for each successive span of three, six, nine, and more months is also shown. A positive correlation between partial and full amplitude can be observed throughout. Three months

CHART 16

RECESSION ANALYSIS,
NONAGRICULTURAL EMPLOYMENT,
PERCENTAGE CHANGE FROM BUSINESS CYCLE PEAKS,
SIMPLIFIED PRESENTATION, 1929–62

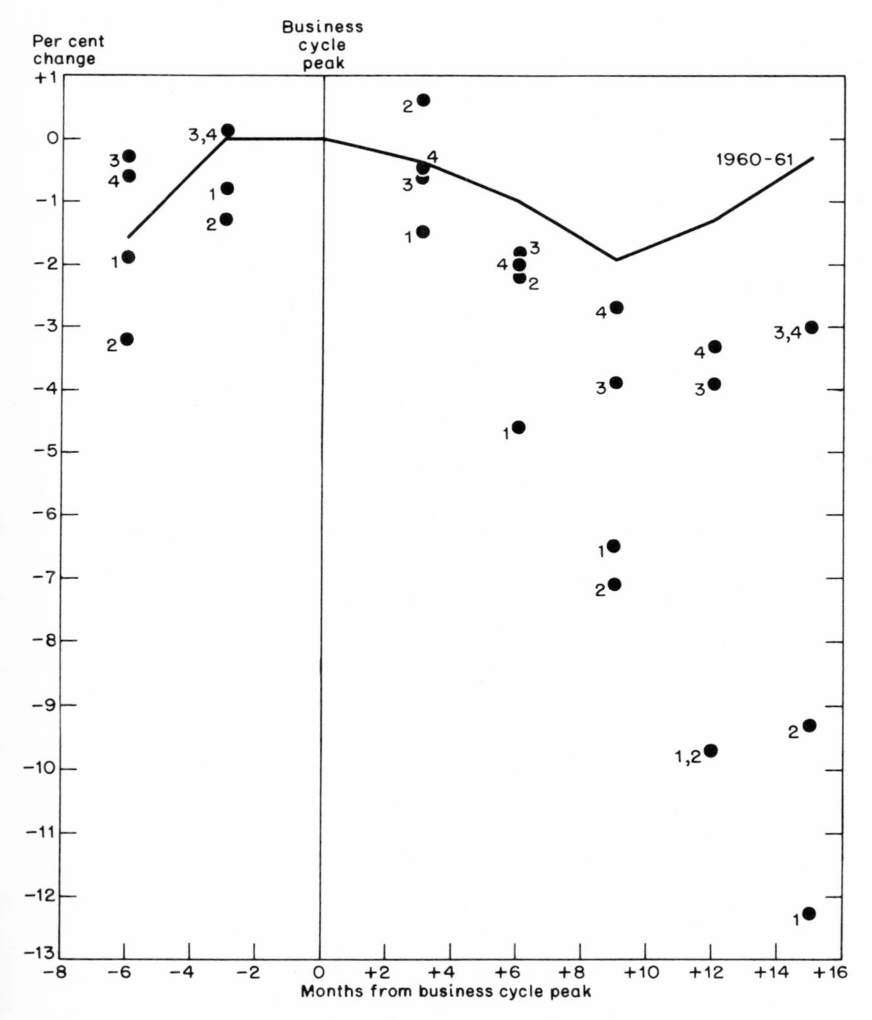

Note: For the recession beginning in 1929, the symbol is 1; for 1937, 2; 1957, 3; 1953, 4; and for 1960, the solid line.

after the business cycle peak, the correlation is relatively weak (+.33); after six months, however, it becomes strong enough to serve as a basis for analysis and anticipation (+.85). The nine-month span shows a temporary decrease of the correlation (to +.61), but thereafter the coefficient increases, and reaches +.96 thirty months after the business cycle peak. While the general feasibility of anticipating the approximate severity of contractions during their initial phase seems supported by these results, there are reasons for raising some questions. As can be seen from Table 11 and Chart 14, most of the included contractions lasted only about a year, yet increasing correlation coefficients are obtained as the measurement period is extended to thirty months. Moreover, one of the contractions, the Great Depression, is so severe that it may have dominated the results during the greater part of the measurement period. It may be well, therefore, to use measures of correlation that are less affected by extreme values, to wit, measures of rank correlation.

Table 13 shows ranks of the partial and eventual recession amplitudes given in Table 11. The corresponding rank correlation coefficients are reported, together with the Pearsonian correlation coefficients, in the last two lines of the table. Note that the rank correlation increases rapidly and after twelve months comes close to unity (+.99) —a coefficient that becomes less astonishing if one realizes that six of the seven contractions occurring during the time period covered lasted about a year, with a range extending from eight to thirteen months. Changes in employment over spans of more than a year show gradually decreasing rank correlation with full contraction amplitudes, reflecting the fact that these changes are more and more affected by subsequent recoveries. The most interesting aspect of the table is the widely divergent behavior of the two types of correlation coefficients. The rank correlation becomes almost perfect after one year and tapers off to a mere .4 before the end of the second year. By contrast, the Pearsonian coefficient is less than .8 after one year, but gradually increases well beyond .9 thereafter. Chart 17 facilitates the understanding of these drastic differences. Its upper panel shows the scatter of partial vs. total amplitudes for a twelve-month span, a span roughly corresponding to the median duration of the included recessions. Hence, most of the twelve-month changes and their ranks are very similar to those of the total amplitudes. However, on the left side, where magnitudes of percentage changes are presented, the enormous eventual

TABLE 13

RANK CORRELATIONS BETWEEN PARTIAL AND FULL AMPLITUDES, NONAGRICULTURAL EMPLOYMENT, BUSINESS CYCLE RECESSIONS, 1929–65

Business Cycle Contractions[a]	*Total Amplitude*		*Ranks of Partial Amplitudes (months after peak)*									
	Per Cent	*Rank*	*3*	*6*	*9*	*12*	*15*	*18*	*21*	*24*	*27*	*30*
Aug. 1929 – Mar. 1933	−30.7	1	1.5	1	3	1.5	1	1	1	1	1	1
May 1937 – June 1938	−10.0	2	7	4	1	1.5	2	2	2	2	2	2
Feb. 1945 – Oct. 1945	−7.6	3	1.5	2	2	3	6	7	7	7	6	6
Nov. 1948 – Oct. 1949	−4.1	4	3	3	5	4	3	5	6	6	7	7
July 1957 – Apr. 1958	−3.8	5	4	6	4	5	4.5	4	4	4	3	3
July 1953 – Aug. 1954	−3.3	6	5	5	6	6	4.5	3	3	3	4	5
May 1960 – Feb. 1961	−1.8	7	6	7	7	7	7	6	5	5	5	4
CORRELATION COEFFICIENTS, PARTIAL VS. TOTAL AMPLITUDES, ALL CONTRACTIONS												
Rank correlation coefficients			+0.44	+0.86	+0.86	+0.99	+0.78	+0.50	+0.39	+0.39	+0.46	+0.43
Pearsonian correlation coefficients			+0.33	+0.85	+0.61	+0.78	+0.85	+0.91	+0.92	+0.94	+0.95	+0.96

Source: Table 11.

[a] In order of contraction amplitudes.

CHART 17

PARTIAL AMPLITUDES *VS*. TOTAL AMPLITUDES, PERCENTAGE CHANGE AND RANK, NONAGRICULTURAL EMPLOYMENT, BUSINESS CYCLE RECESSIONS, 1929–61

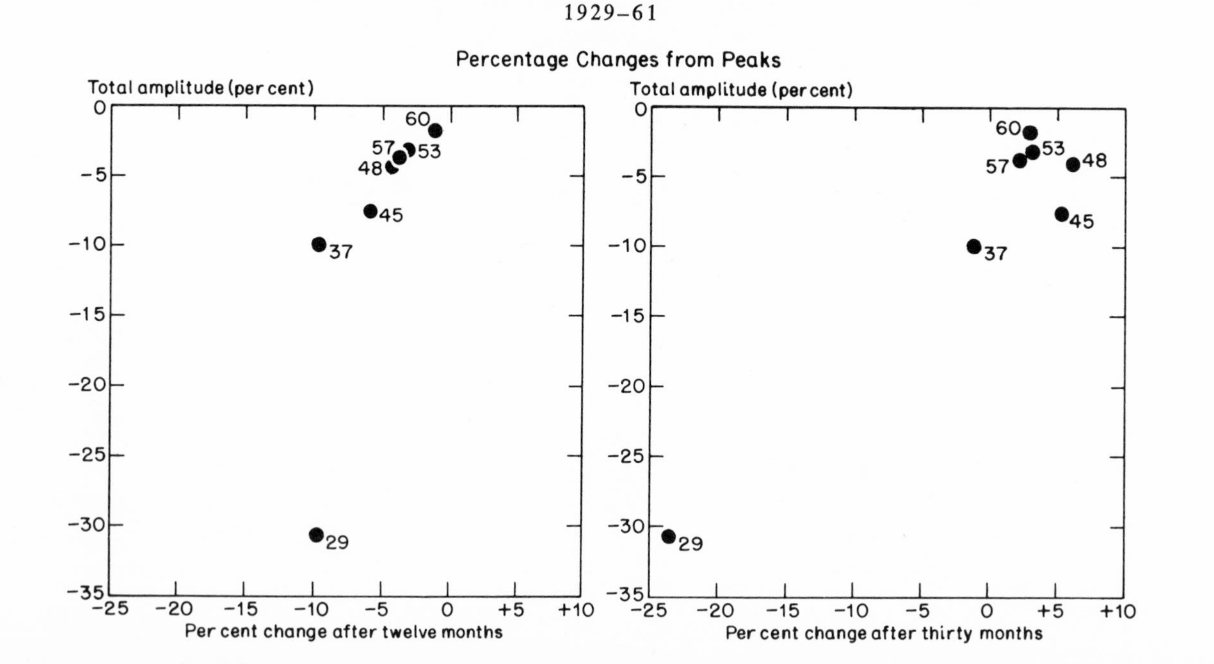

CHART 17

(Concluded)

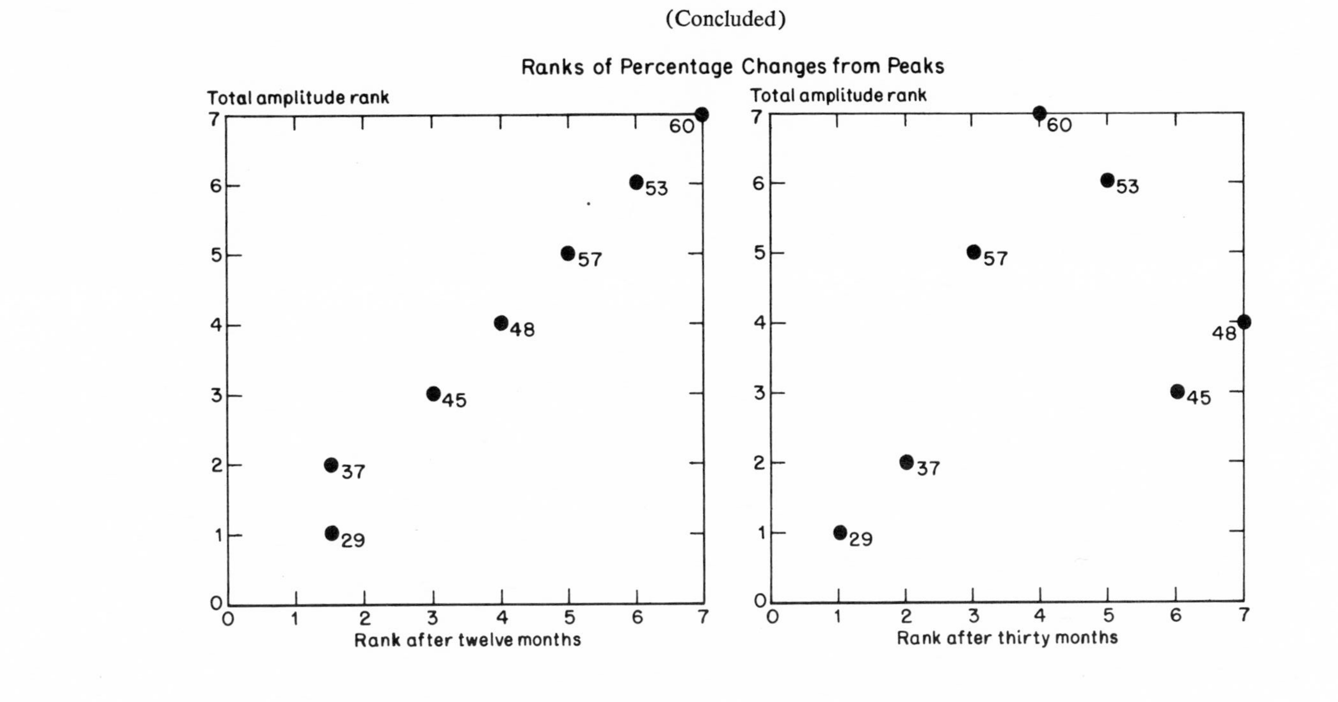

decline of employment during the Great Depression leads to an extreme, nonaligned observation (marked 29–33). It is this observation that limits the Pearsonian correlation coefficient to .8. The ranks, which are not affected by measured extremes, are shown on the right upper panel. They are almost perfectly aligned, except for one tie. This situation is described by the rank correlation coefficient of .99. The picture is markedly different for the thirty-month span, by the end of which most of the included recessions had ended and employment was on the ascent. The lower panels of the chart illustrate the situation. Here, the measured percentage deviations from the preceding peak are small, loosely assorted, and closely bunched for the milder recessions; but the severe declines of the interwar period determine a steep regression line from which the deviations are relatively small, hence the high Pearsonian correlation coefficient of .96. The ranks, by contrast, reflect the haphazard order of the mild deviations (under 10 per cent) and thus lead to a rank correlation coefficient of merely +.43. This demonstrates that summary measures can often be opaque and even misleading if the underlying structure is not analyzed. Comparisons between alternative measures are often highly beneficial and instructive. They not only prevent rash conclusions from either measure but help to elucidate the processes under review. And this elucidation can frequently be obtained at very low incremental costs, once alternative approaches are available in programmed form.

A strong correlation seems to exist between the durations and the amplitudes of recessions. This should be of considerable interest to those who wish to use recession analysis as a forecasting tool. In the present example of nonagricultural employment, the Pearsonian coefficient of correlation between the duration of business cycle contractions and the percentage changes from peaks is as high as −.96. As can be seen in the left panel of Chart 18, the 1929–33 contraction was longest and deepest, that of 1937–38 was second in both respects, and the postwar contractions were shortest and mildest. However, the high correlation coefficient is unduly influenced by the extremely long and deep contraction after 1929. No correlation between durations and amplitudes can be discovered among the mild postwar contractions—both because of their common mildness and because of the increasing role of governmental interference.

The recession-recovery program, in its present form, does not rank percentage changes and total phase amplitudes. Therefore, users in-

CHART 18

PHASE AMPLITUDES *VS.* PHASE DURATIONS, NONAGRICULTURAL EMPLOYMENT, 1929–61

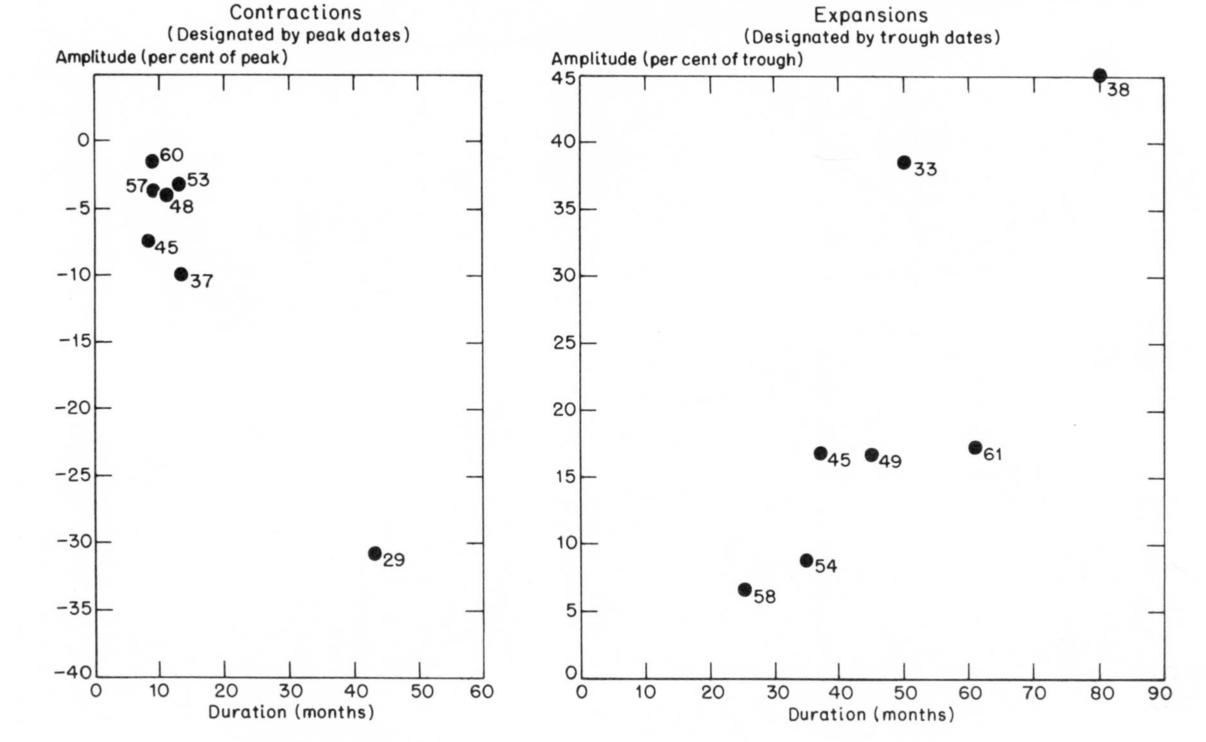

Source: Tables 7 and 8.

terested in the degree to which the ranking of the phase amplitudes is approximated by the ranking of the partial amplitudes for various time intervals from the peak must derive the ranks from the appropriate tables. The degree of correlation between partial and eventual amplitude should be helpful in attempts to gauge the prospective severity of a current contraction. The described procedure can be carried through easily if rank correlations are to be established for only a few selected spans and for a small number of series. When it is desirable to find the span of highest correlation among many spans for a considerable number of series, a programmed approach becomes clearly preferable.[6] The same is true for Pearsonian correlation coefficients.

In order to illustrate the broad usefulness of this technique for analysis and forecasting of business conditions, let us quote some generalizations, which Geoffrey Moore derived on the basis of its application to many time series.

1. When a business recession begins, most broad indicators of aggregate economic activity (production, employment, income, trade) show relatively slight declines, and during the first six months of the recession the magnitude of the declines bears little relation to the ultimate severity or depth of the recession.

2. About six months after a recession begins, the percentage declines from peak month to the current month in most economic aggregates are smaller in mild recessions than in severe recessions, and this ranking is maintained in succeeding months with little change.

3. When such comparisons are made for types of economic data that typically begin declining before a recession starts (for example, new orders, construction contracts, the average workweek, stock prices) the distinction between mild and severe recessions begins to appear as early as three or four months after the recession begins, and is also substantially maintained in succeeding months.

4. Although frequently both mild and sharp business contractions have ended within about a year, the recovery to the previous peak level has been accomplished much more quickly after mild contractions. Hence the period of depressed activity has been much longer when the contraction proceeded at a rapid rate.

[6] Ranking of changes (for various time spans) and of total phase amplitudes can be added to the program, if demanded by users. Also, rank correlation coefficients and Pearsonian correlation coefficients can be provided. The correlation coefficients used in the present paper were calculated on electronic computers, but with the help of separate programs.

5. While the above conclusions suggest that a rough ordering of recessions according to severity can be made within four to six months after the onset, they do not imply that either the ultimate depth or the duration of recessions can be reliably forecast by this means. Many factors not taken into account by the method, such as governmental measures taken to combat depression, have an important bearing on the severity and duration of business contractions. The method appears useful primarily in providing a yardstick against which a current decline in various aspects of economic activity can be gauged, thereby facilitating a more accurate and enlightened appraisal of what has already taken place. This in itself might facilitate the development of appropriate counter-cyclical programs.

6. Measures of the strength of various counter-cyclical factors (for example, unemployment compensation payments, increased governmental expenditures, easier credit terms, lower taxes) at similar stages of recession might be developed on the same plan as described here. . . . Such measures might be of assistance in judging the prospects for further business contraction or for a resumption of economic expansion.

7. Several months before a recession comes to an end and an upturn in aggregate activity occurs, a progressive narrowing of the scope of contraction ordinarily becomes visible. Fewer activities continue to decline, more begin to rise. It appears first in series of the "leading" type. The more extensive and more sustained this reduction in the scope of the contraction is, the more likely that it marks the real end of recession rather than an abortive recovery. Information of this sort may help to identify an upturn in aggregate activity at about the time it occurs or shortly thereafter.[7]

RECOVERY PATTERNS

Recoveries can be analyzed in much the same way as recessions. That is, one can measure the percentage changes of individual time series from cyclical trough levels over spans of increasing length and observe how a current expansion fares in comparison with preceding ones. This procedure was illustrated in Table 12 and Chart 15. Note that the percentage rises of employment during the early months were closely related to their eventual amplitudes during the expansions. After three or four months the 1933–37, 1938–45, and 1945–48 employment expansions began to emerge as relatively vigorous, the 1954–57 and 1958–60 expansions as mild. That this compares fairly well with the

[7] Moore, *Measuring Recessions,* p. 264. This paper is reprinted in Moore (ed.), *Business Cycle Indicators,* Vol. I, Chapter 5. See also Appendix C to that volume.

eventual amplitudes can be seen in the last lines of Tables 12 and 14 which present Pearsonian correlation coefficients and rank correlation coefficients describing the relationship between partial and full amplitudes of expansions in employment, during business cycle expansions.

The rank correlation coefficients during the diagnostically crucial period of six to twelve months after the turn were markedly lower than those for comparable contraction spans. This is the case whether the measures are computed for all recoveries after 1933, or whether the currently incomplete expansion, starting with 1961, is left out. The comprehensive rank correlation coefficient reached +.64 six months after the trough. However, this level is deceptive, since thereafter it declines to +.44 and +.36. The Pearsonian coefficients are, except for the three-month interval, a bit higher than the rank coefficients; they also are appreciably lower than the Pearsonian coefficients for contractions (given in Table 13). The differences between the recession and the recovery relationships can perhaps be best summarized in schematic form. Some of the described relationships apply only to the particular activity examined. However, the lower correlation between partial and total expansion amplitudes, compared with recession amplitudes, tends to predominate widely.

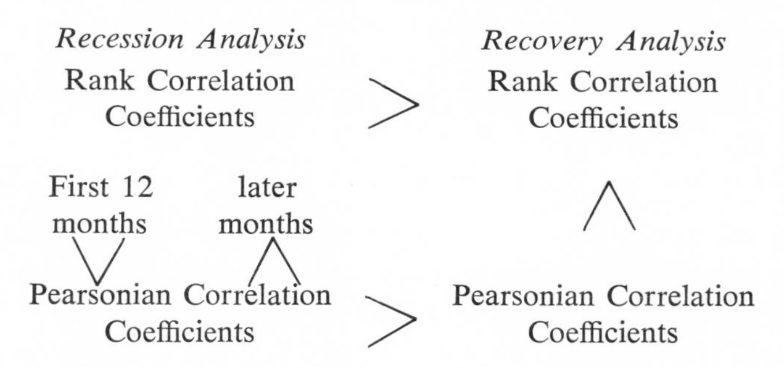

The relationship between durations and amplitudes is less close in expansions than in recessions. For nonagricultural employment, the Pearsonian coefficient of correlation between the durations of business cycle expansions and employment amplitudes during the same period is +.81, as compared to the corresponding measure of −.96 for contractions. However, the correlation for expansions is more pervasive and less affected by extreme observations, as can be seen in the right panel of Chart 18. Since expansion rates during the mild postwar

TABLE 14

RANK CORRELATIONS BETWEEN PARTIAL AND FULL AMPLITUDES (MEASURED FROM TROUGHS), NONAGRICULTURAL EMPLOYMENT, BUSINESS CYCLE EXPANSIONS, 1933–66

Business Cycle Expansions[a]	*Total Amplitudes*		*Ranks of Partial Amplitudes (months after trough)*									
	Per Cent	*Rank*	*3*	*6*	*9*	*12*	*15*	*18*	*21*	*24*	*27*	*30*
June 1938 – Feb. 1945	45.3	1	3	3	3.5	4	4	4	4	3	3	2
Mar. 1933 – May 1937	38.5	2	1	1	1	1	1	1	1	1	1	1
Feb. 1961 – Mar. 1966[b]	17.2	3	6	6	7	7	7	7	7	7	7	6
Oct. 1945 – Nov. 1948	16.9	4	2	2	2	2	2	2	2	2	2	3
Oct. 1949 – July 1953	16.6	5	5	4	3.5	3	3	3	3	4	4	4
Aug. 1954 – July 1957	8.8	6	4	5	5	5	5	5	5	5	5	5
Apr. 1958 – May 1960	6.7	7	7	7	6	6	6	6	6	6	6	7
CORRELATION COEFFICIENTS, PARTIAL VS. TOTAL AMPLITUDES												
Rank correlation coefficients for all expansions			+0.61	+0.64	+0.44	+0.36	+0.36	+0.36	+0.36	+0.50	+0.50	+0.75
Rank correlation coefficients for expansions excl. 1961–66			+0.77	+0.83	+0.76	+0.66	+0.66	+0.66	+0.66	+0.83	+0.83	+0.94
Pearsonian correlation coefficients for all expansions			+0.57	+0.66	+0.52	+0.46	+0.54	+0.58	+0.55	+0.58	+0.72	+0.86

Source: Table 12.

[a] In order of expansion amplitudes.

[b] Last month included.

cycles were observed to be closely clustered, it would follow that expansions with longer duration also tend to have larger amplitudes. This relationship is closer than that between amplitudes of expansion and amplitudes of preceding contractions, or that between total and partial expansion amplitudes.

The relatively low correlation between partial and full expansion amplitudes implies that the latter cannot be very successfully anticipated by measuring the vigor during their early stages whether the expansions occur in individual series or in economic activity at large. This is, however, no reason to despair of the possibility of anticipating business conditions during recovery periods. After all, the attempt to anticipate amplitudes of expansions that may last five years or more by observing them during the first six months or so should be regarded prima facie as an unreasonably optimistic endeavor. By shortening the forecasting period and modifying the approach, some meaningful generalizations can be developed about the process of economic recovery from cyclical declines.

Let the expansion period be divided into two segments, the portion until a given activity reaches preceding peak levels (recovery segment) and that from these recovery levels to the next peak (growth segment), and then concentrate on the first portion. Furthermore, let recovery levels be expressed in terms of the peak preceding the recovery, rather than the initial trough. The procedure is illustrated in Table 15 and in Chart 17. In this chart, the vertical axis measures the deviation of the series from the preceding reference *peak* levels in percentage of these peaks.[8] The horizontal axis is calibrated in months, measuring increasing spans from the respective trough months. Table 15 and, perhaps more effectively, Table 16 (which removes the percentage base bias caused by differential trough levels) show that employment expansions tended to be more vigorous after severe contractions. If the amplitude of preceding contractions was large (last column), employment tended to increase more sharply during the first two years or so. Table 16 shows a pronounced tendency for the recovery percentages, at any given month, to be higher after contractions of severe amplitudes. However, after severe contractions it took longer to regain previous peak levels (recovery levels) than after mild con-

[8] Here the series cannot be expected to converge at the initial trough of the recovery, as is the case when the trough level itself is made the base for the calculations. The differences in the levels of the starting point reflect, of course, the differing severities of the preceding declines.

TABLE 15

RECOVERY ANALYSIS, NONAGRICULTURAL EMPLOYMENT, PERCENTAGE CHANGE FROM PRECEDING BUSINESS CYCLE PEAK LEVELS, 1932–64

Date of Trough[a]	*Standing at Previous Peak*[b] *(thousands)*	*Percentage Change (months before trough)*		*Percentage Change at Trough*	*Percentage Change (months after trough)*										*Amplitude of Preceding Contraction*[b]
		6	*3*		*3*	*6*	*9*	*12*	*15*	*18*	*21*	*24*	*27*	*30*	
Mar. 1933	33,222	−29.2	−29.1	−31.3	−28.1	−22.9	−22.3	−19.7	−18.6	−20.3	−18.7	−17.0	−16.7	−15.3	−30.7
June 1938	31,904	−4.4	−7.9	−10.2	−8.6	−6.2	−5.5	−4.7	−2.4	−1.0	−1.0	0	+3.0	+7.0	−10.0
Oct. 1945	41,740	−1.0	−2.7	−7.8	−4.8	−2.2	+0.5	+3.0	+4.2	+4.2	+4.6	+6.1	+7.0	+6.2	−7.6
Oct. 1949	45,077	−2.3	−3.6	−5.0	−3.6	−1.8	+0.7	+3.2	+4.9	+6.2	+6.4	+6.1	+7.1	+7.6	−4.1
Apr. 1958	53,011	−0.6	−1.8	−3.9	−3.9	−3.0	−1.1	+0.5	+1.3	+0.4	+2.2	+2.9	+2.3	+1.8	−3.8
Aug. 1954	50,378	−2.1	−3.0	−3.3	−2.5	−1.6	+0.1	+1.1	+2.3	+3.5	+4.0	+4.1	+4.7	+5.3	−3.3
Feb. 1961	54,407	−0.4	−1.0	−1.9	−1.3	−0.3	+0.4	+1.1	+2.0	+2.4	+2.7	+3.0	+3.8	+4.3	−1.8
Average		−5.7	−7.0	−9.1	−7.5	−5.4	−3.9	−2.2	−0.9	−0.7	0	+0.7	+1.6	+2.4	−8.8
Average deviation		6.7	6.6	6.7	6.2	5.2	5.7	5.7	5.5	5.7	5.6	5.3	5.2	5.2	6.6

Source: Output Tables 3A-3 and 3A-8.

[a] In order of amplitudes of preceding contractions (see last column).

[b] Three-month average centered at turn.

TABLE 16

RECOVERY ANALYSIS, NONAGRICULTURAL EMPLOYMENT, CHANGE FROM BUSINESS CYCLE TROUGHS AS A PERCENTAGE OF LEVELS AT PRECEDING BUSINESS CYCLE PEAKS, 1932–64

Date of Trough[a]	*Standing at Trough (thousands)*	*Standing at Previous Peak (thousands)*	*Percentage Change (months before trough)*		*Percentage Change (months after trough)*										*Amplitude of Preceding Contraction*[b]
			6	*3*	*3*	*6*	*9*	*12*	*15*	*18*	*21*	*24*	*27*	*30*	
Mar. 1933	23,030	33,222	+1.5	+1.5	+2.6	+7.8	+8.4	+11.0	+12.1	+10.4	+12.0	+13.7	+13.9	15.4	−30.7
June 1938	28,725	31,904	+5.6	+2.9	+1.4	+3.8	+4.5	+5.3	+7.5	+9.0	+9.0	+10.0	+13.0	17.0	−10.0
Oct. 1945	38,559	41,740	+6.7	+4.9	+2.8	+5.4	+8.2	+10.7	+11.8	+11.8	+12.2	+13.7	+14.6	13.8	−7.6
Oct. 1949	43,215	45,077	+1.9	+0.5	+0.6	+2.4	+4.8	+7.3	+9.0	+10.4	+10.5	+10.2	+11.2	11.7	−4.1
Apr. 1958	50,978	53,011	+1.1	+0.3	+0.8	+1.6	+3.3	+4.2	+5.3	+6.4	+7.0	+7.1	+7.6	8.2	−3.8
Aug. 1954	48,720	50,378	+3.4	+2.1	0.0	+0.9	+2.9	+4.5	+5.4	+4.5	+6.4	+7.1	+6.4	6.0	−3.3
Feb. 1961	53,451	54,407	+1.4	+0.7	+0.4	+1.4	+2.1	+2.8	+3.7	+4.1	+4.5	+4.7	+5.6	6.0	−1.8

Source: Computed from Output Table 3A-5.

[a] In order of amplitudes of preceding contractions (see last column).

[b] Three-month average centered at turn.

tractions. This appears clearly in Chart 19. One or two years after the trough the levels of employment were still, by and large, in an order similar to that of the amplitudes of the preceding contractions. Table 15 contains averages and average deviations for the measured percentage changes. Note that, over the included cycles, previous peak

CHART 19

RECOVERY ANALYSIS, NONAGRICULTURAL EMPLOYMENT, PERCENTAGE CHANGE FROM PRECEDING BUSINESS CYCLE PEAK, 1932–63

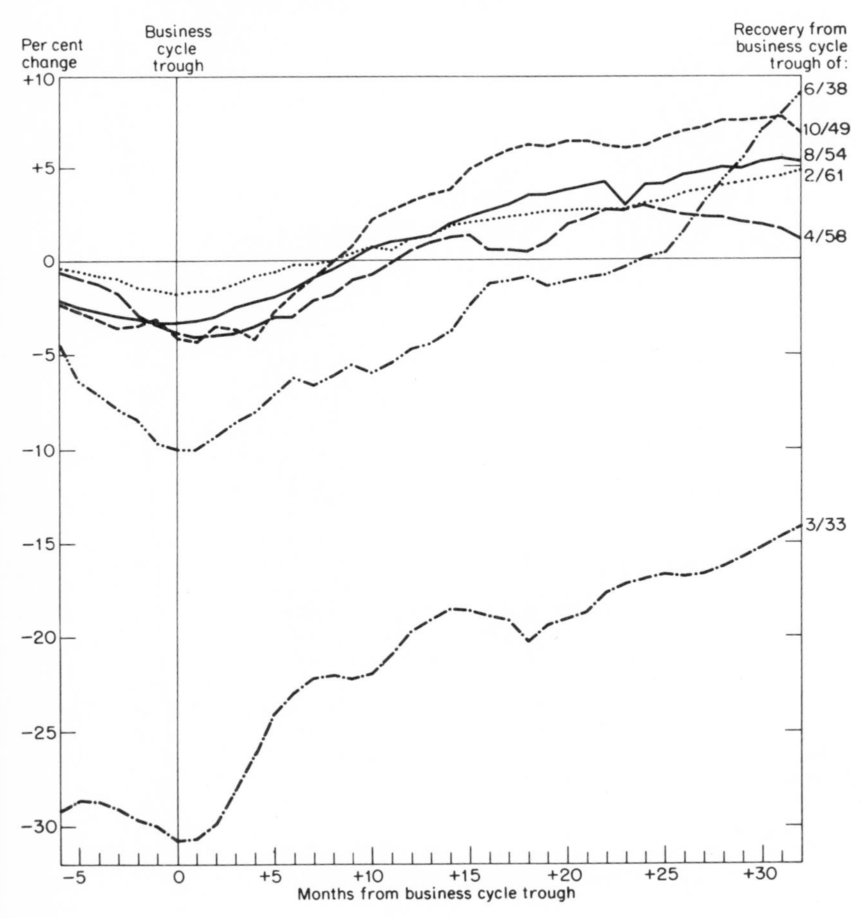

levels were reached after about twenty-one months, on the average. The average deviation of the relative levels showed a mild tendency to decrease—in keeping with the observation that the vigor of expansions is inversely related to the depth of the preceding decline.

If the time span that contains a contraction and the subsequent recovery to the preceding peak level[9] is designated as a period of "depressed activity," some further characteristics of cyclical recoveries in employment may be described. Table 17 shows durations and amplitudes of employment cycles in terms of contraction, recovery, depressed activity, growth, and total expansion—all based on employment levels at initial business cycle peaks. The length of the period of depressed activity is correlated with the severity of the initial decline; this correlation is indeed closer than that between the duration and the severity of the decline itself. By contrast, the amplitude and duration of the growth segment are less closely related to the preceding decline. There exists, historically, some inverse relationship between the depth of a contraction and the amplitude of the growth segment of the following expansion. After the deepest contraction (1929–33) included in the sample, the amplitude of the growth segment was smallest; after the mildest contraction (1960–61), the growth amplitude was largest;[10] after contractions of intermediate severity, the growth was moderate. However, within the middle group, covering four recessions, the inverse correlation does not prevail, perhaps because the differences between recession amplitudes are so small. The broad inverse relationship suggests that stability breeds stability: The absence of violent downward adjustments helps to pre-

[9] The measure of recovery to past peak levels should be determined in such a way that the result is cyclically significant and not due to erratic movements. A three-month moving average was used to establish the termination of the recovery phase; that is, the recovery phase was regarded as concluded when a three-month average of the series reached or exceeded the previous peak standing, also a three-month average. For more erratic series a longer moving average might be desirable. It should be noted that the recovery to previous peak employment levels is not a measure that fits precisely into the reference analysis framework. The relevant recovery measure, in that framework, would relate to employment levels reached when *general business activity* regained previous peak levels. Since these dates are not established and since their establishment lies beyond the scope of the present study, the measure described above was used.

[10] This ignores the growth experience that includes the expansion during World War II—an expansion that could not possibly be related to the 1937–38 decline in business activity.

TABLE 17

CHARACTERISTICS OF RECOVERIES RELATED TO PRECEDING RECESSIONS, NONAGRICULTURAL EMPLOYMENT, REFERENCE CYCLE ANALYSIS, 1929–66

Date of				Duration (in months)					Amplitudes (in per cent of initial peak levels): Total				Amplitudes: Per Month			
Initial Peak	*Trough*	*Reaching Previous Peak Level*	*Terminal Peak*	*Contraction*	*Recovery*	*Depressed Activity*	*Growth Segment*	*Total Expansion*	*Contraction*	*Recovery*	*Growth Segment*	*Total Expansion*	*Contraction*	*Recovery*	*Growth Segment*	*Total Expansion*
8/29	3/33	10/40	5/37	43	91	134	—	50	−30.7	+30.8	−4.1	26.7	−0.71	0.34	—	0.53
5/37	6/38	7/40	2/45	13	25	38	55	80	−10.0	+10.6	30.2	40.8	−0.77	0.42	0.55	0.51
2/45	10/45	7/46	11/48	8	9	17	28	37	−7.6	+8.3	7.3	15.6	−0.95	0.92	0.26	0.42
11/48	10/49	7/50	7/53	11	9	20	36	45	−4.1	+5.0	10.9	15.9	−0.37	0.56	0.30	0.35
7/57	4/58	4/59	5/60	9	12	21	13	25	−3.8	+4.2	2.3	6.5	−0.42	0.35	0.18	0.26
7/53	8/54	5/55	7/57	13	9	22	26	35	−3.3	+3.4	5.1	8.5	−0.25	0.38	0.20	0.24
5/60	2/61	11/61	3/66[a]	9	9	18	52	61	−1.8	+2.0	15.2	17.2	−0.20	0.22	0.29	0.28
ALL CYCLES																
Average				15.1	23.4	38.6	—	47.6	−8.76	9.19	9.56	18.74	−0.524	0.456	—	0.370
Average deviation				7.9	19.7	27.3	—	13.8	6.63	6.58	7.89	8.57	0.245	0.163	—	0.100
Weighted average													−0.576	0.392	—	0.392
ALL CYCLES EXCL. 1929–37																
Average				10.5	12.2	22.7	35.0	47.2	−5.10	5.58	11.83	17.42	−0.493	0.475	0.297	0.343
Average deviation				1.8	4.3	5.1	12.7	15.5	2.47	2.58	7.24	7.80	0.244	0.177	0.086	0.084
Weighted average													−0.484	0.458	0.338	0.368

[a] Last month included.

vent crass distortions and subsequent overcorrections; it tends to improve foresight, engender confidence, discourage speculative excesses, and facilitate managerial and governmental guidance. While the described tendencies are not reliable enough to form a safe basis for anticipations, they provide another example of the manifold advantages of cyclical stability.[11]

From all this it follows that the correlation between total expansion and the preceding contraction tends to be high when the recovery forms a substantial portion of the total expansion (as often occurs after deep contractions), and low if the portion is small (as in the expansion that started in 1961). Recognition of these relationships helps in the interpretation of recovery patterns.

The observed regularities transcend the recovery patterns of nonagricultural employment and may indeed characterize cyclical behavior in general.[12] Recovery analysis, carried out by Geoffrey Moore for employment, output, profits, and stock prices, led to the following generalizations:

1. Recoveries in output, employment, and profits have usually been faster [i.e., growth rates have been higher] after severe depressions than after mild contractions.
2. Despite the faster pace after severe contractions, recovery to the previous peak level has taken longer when the preceding contraction has been severe.
3. Nearly every business expansion has carried total output, employment, and profits beyond the level reached at the preceding peak.
4. The rate of growth in output, employment, and profits has usually been largest at the initial stages of a business expansion. Thereafter, slower growth has been the rule, especially after the preceding peak level has been regained.
5. Stock prices, unlike output, employment, or profits, have advanced more rapidly after mild recessions than after severe contractions.[13]

[11] Geoffrey H. Moore called attention to the inverse relationship between contraction amplitudes and subsequent growth in "Business Indicators—What They Tell Us," a paper presented at the *Tenth Annual Conference on the Economic Outlook,* University of Michigan, 1962.

[12] See also Chart 6 and the related comments in Chapter 3.

[13] Geoffrey H. Moore, "Leading and Confirming Indicators of General Business Changes," in Moore (ed.), *Business Cycle Indicators,* Vol. I, Chapter 3, p. 92. This chapter, particularly the section "Measuring the Vigor of a Business Recovery," is important for users of recovery analysis.

The broad usefulness of recovery patterns for current business conditions analysis has been succinctly described as follows: "The method can be used to appraise a business cycle recovery month by month as it develops; to measure its vigor, scope, and unusual features; to derive some rough notion of its probable course and duration and to check the reasonableness of forecasts derived by other means, always remembering that typical rates of recovery and patterns of change vary from one economic activity to another."[14]

Although the generalizations cited above refer largely to the recovery segment, the values of comparative analysis are not restricted to this portion of the total expansion. Prior expansions (including their growth phases) can be used as a "grid" against which the behavior of a current expansion can be judged, at least until it outlasts the duration of previous expansions. Since the contraction of 1960–61 was short and mild, previous peak levels were reached by many activities early in the expansion—in the case of nonagricultural employment, indeed, before the end of the year 1961. Recovery analysis along the described lines extends, of course, beyond that date and permits identification of the major characteristics of the following expansion.[15]

VARIANTS OF USES AND APPROACHES

NONFORECASTING USES

The recession-recovery analysis technique lends itself to applications other than the rough classification of current expansions and contractions as mild or strong. As mentioned earlier, it can be used to bring out the salient qualitative characteristics of a current cycle phase. The analysis may, for example, show that prices have risen sharply but production and employment only mildly, in contrast to some earlier

[14] *Ibid.*, p. 88.

[15] For selected business cycle indicators, recovery analysis is regularly performed and published in the monthly periodical *Business Conditions Digest* (formerly *Business Cycle Developments*), U.S. Department of Commerce, Bureau of the Census. During an expansion period in general business conditions, use is made of recovery analysis; during a contraction period, the tables and charts reflect the recession analysis approach. Use of recovery analysis for the characterization of the expansion starting February 1961 was made in Julius Shiskin's articles published in the January 1965 issue of *Business Cycle Developments,* and in the January 1970 issue of *Business Conditions Digest.*

expansion in which a different, even the reverse, situation may have prevailed.

Second, the analysis can be used to classify and characterize historical rather than current expansions and contractions; that is, it can be used for historical analysis, as a supplement or alternative to business cycle analysis. In Chart 20, the reduction of unemployment dur-

CHART 20
RECOVERY ANALYSIS, UNEMPLOYMENT RATE, PERCENTAGE CHANGE FROM PRECEDING BUSINESS CYCLE PEAK, 1933–63

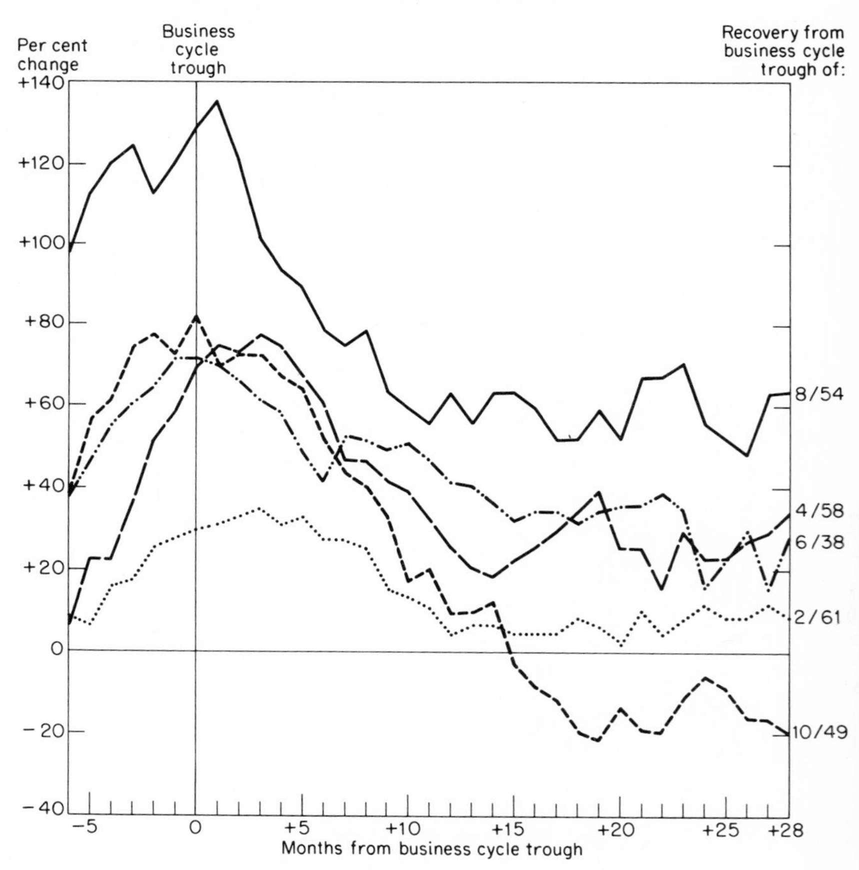

ing business cycle expansions is shown relative to unemployment levels at preceding business cycle peaks (zero levels on chart). Following the 1948–49 contraction, the unemployment rate was reduced to previous prosperity levels after about fifteen months, that is, by January 1951. In none of the other postwar expansions did the reduction of unemployment lead to the rate prevailing at the preceding business cycle peaks within the twenty-eight months depicted in the chart. That is, none of the lines (except that starting in 1949) ever touched the zero line, and during the three last recoveries the degree to which previous prosperity unemployment was approximated (within two and a half years) varied with the height of the unemployment rate during the preceding contraction, as reflected by the position of the lines at the business cycle trough. In general, unemployment rates show fast declines after business cycle troughs for a period of a year or a year and a half; thereafter they tend to decrease only mildly or maintain their levels. Note also that there is an historical sequence in the amplitudes of unemployment during business cycle contractions: The contractions gradually became milder. However, these remarks are all based on changes in relatives, that is, they do not consider whether the unemployment rate was high or low in terms of the labor force, and what the changes were in these terms. If a user finds this analysis inadequate, a variant based on original units should be selected (see the next section). Use of the recession-recovery approach for the historical analysis of business cycles is important since it is at least possible that certain regularities of cyclical behavior may be as effectively or more effectively described in terms of conventional chronological time than in terms of phase fractions.

Third, the analysis can be used for interregional, interindustry, or other cross-sectional comparisons. This can be done in a variety of ways. One obvious possibility is simply to use the technique to compare, for a given activity, the cyclical changes in, say, different states, so that one may see how a recession in New York compares with that in neighboring states. But the analysis becomes more instructive, and stays at the same time closer to its original design, if cross-classification and historical analysis are combined. One may choose to apply the standard approach to cyclical indicators for states or regions and to observe how the characteristics of a given recession in one state (shown against a grid of earlier recessions) compare with those observable in other states or in the nation as a whole.[16] This would emphasize the regional variations in the historical peculiarities of a given recession. Chart 21 shows, for instance, that employment in Florida increased rapidly during the 1953–54 and 1957–58 recessions, while in Texas, West Virginia, and the United States as a whole, it dropped markedly. During the more recent 1960–61 contraction, by contrast, Florida's employment levels rose only mildly in spite of the fact that this recession brought less substantial declines of employment in the United States or the other two states. Moreover, while employment in West Virginia showed the beginnings of a vigorous recovery after the 1953–54 contraction (a recovery that continued beyond the period included in the chart), there was little recovery from the 1958 and, for a while, from the 1961 trough levels—in spite of the fact that in the nation and in most other states employment rose promptly from recession levels. As the present interest of this study lies in the illustration of various uses of recession-recovery analysis rather than in a discussion of state employment cycles, the above comments on Chart 21 may well suffice. It is possible to go one step further, however, in analyzing regional differences. Employment changes in highly industrialized states have a good deal of family resemblance, which makes it difficult to distinguish between them. Differential behavior can perhaps be brought out of first "adjusting" the activities for the national changes (dividing the relatives for the state by those for the nation) and then applying the analysis.

Finally, recovery analysis and recession analysis may find applications as a tool for market analysis, sales analysis, and similar endeavors. In general recovery analysis, the attempt is usually made to evaluate the vigor of a recovery by comparing a current expansion with past expansions. Similarly, an industry can gauge its recovery—perhaps relative to other industries or to broad industrial aggregates—by observing how its employment, output, prices, profits, etc., fared in a particular expansion as compared to earlier ones. An analogous approach can be used for recession analysis. And the general technique may lend itself to comparative analyses of company and industry performance, if there is sufficient cyclical responsiveness in company operations to make such comparisons meaningful.

[16] For an example of such use, see the authors' *Economic Indicators for New Jersey,* New Jersey Department of Labor and Industry, Division of Employment Security, 1964, Charts A to I.

CHART 21

COMPARATIVE RECESSION ANALYSIS, NONAGRICULTURAL EMPLOYMENT IN TEXAS, FLORIDA, WEST VIRGINIA, AND THE UNITED STATES, PERCENTAGE CHANGE DURING BUSINESS CYCLES, 1952–62

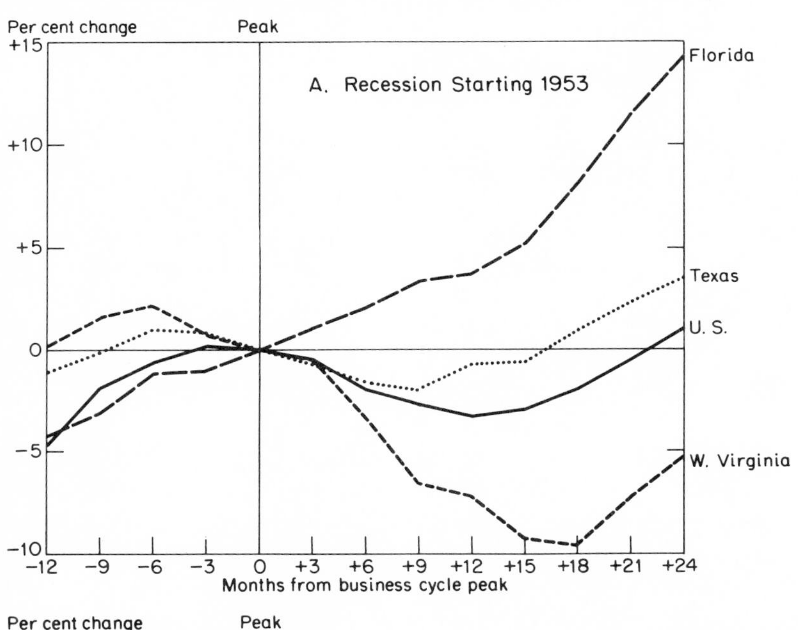

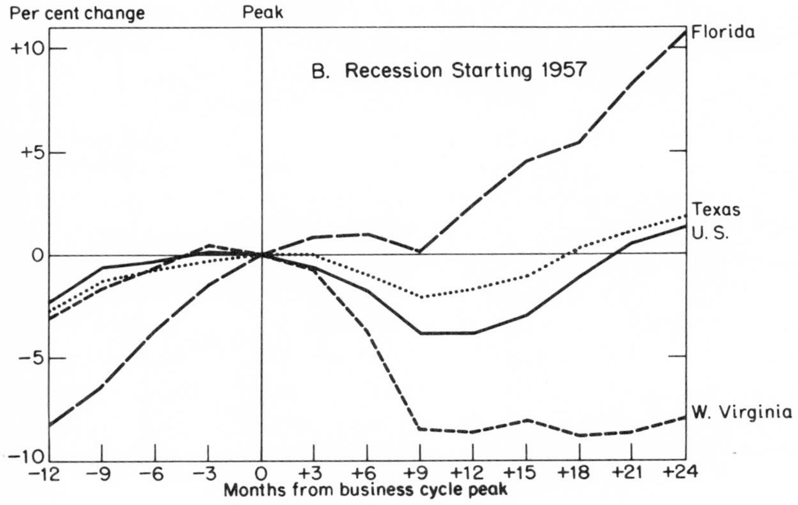

CHART 21

(Concluded)

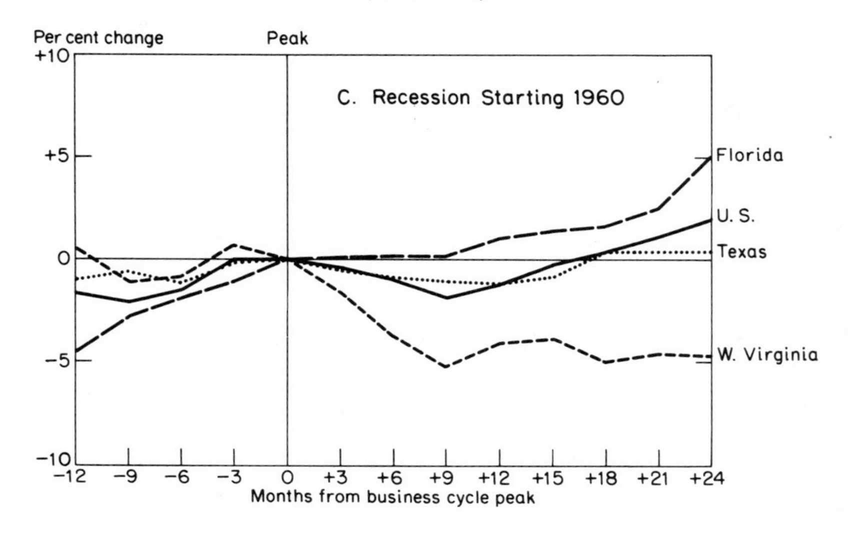

Chart 22 and Table 18 illustrate the application of recovery analysis to company performance; they compare the sales experience of a fairly diversified manufacturing enterprise with that of all manufacturing in the United States. The company representatives were under the impression that the company's sales (after adjustment for mergers, acquisitions, and the like) were remarkably similar to those of manufacturing in general. While a cursory examination might convey this impression—cycles correspond and long-term trends are upward—the more detailed comparison afforded by recession-recovery analysis leads to some modification of this view. It is true that during cyclical recoveries company and industry sales correspond rather well, although company sales tend to turn up earlier and to rise above their previous trough levels, in expansions, by more percentage points than do industry sales (see the two last columns of Table 18). Chart 22 shows that the only case in which drastic divergence between company and industry experience occurs is during 1950–51, when the company responded to the Korean War conditions quite differently than did manufacturing as a whole (second year of panel A). During contraction periods, rather systematic differences exist. The sales experiences of the company contrast favorably with those of manufacturing. Chart

CHART 22
RECOVERY ANALYSIS,
COMPANY SALES AND SALES OF MANUFACTURES,
PERCENTAGE CHANGE
FROM TROUGH IN SALES OF MANUFACTURES, 1949–63

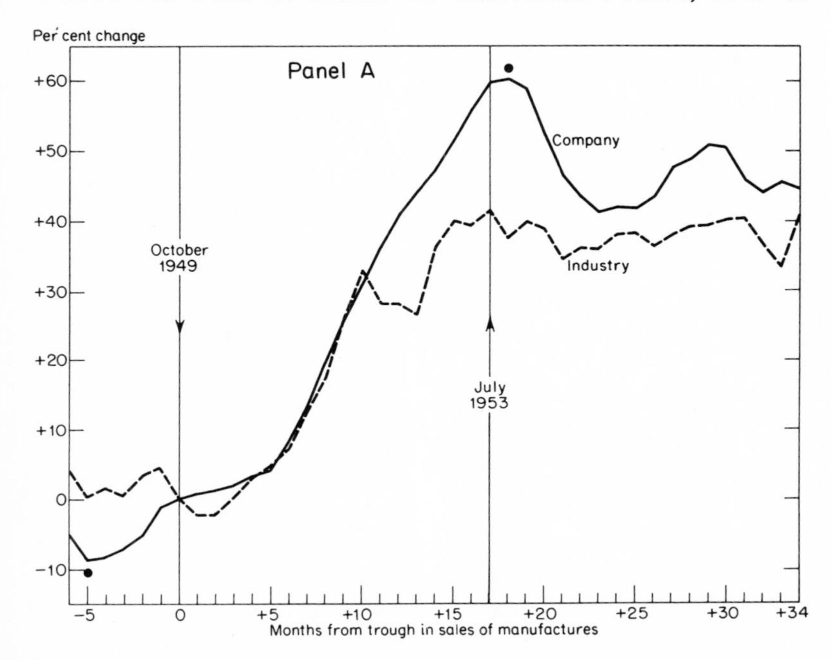

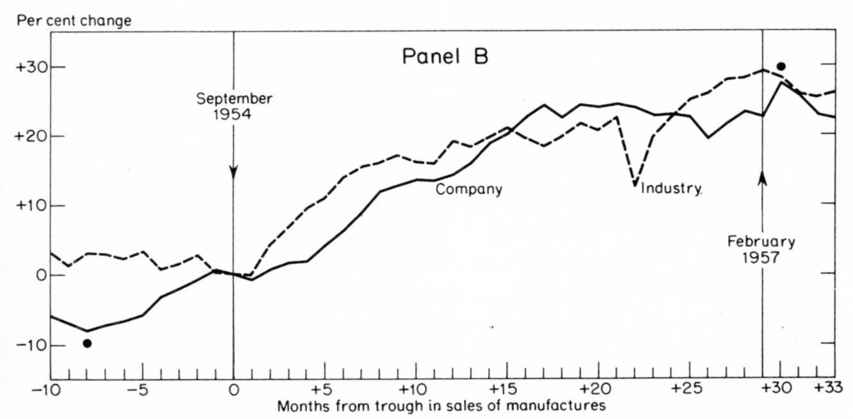

Note: Vertical lines denote months of cyclical turns in sales of manufactures; percentage changes are computed from three months averages centered around the trough (peak). Circles denote cyclical turning points in the company series.

CHART 22

(Concluded)

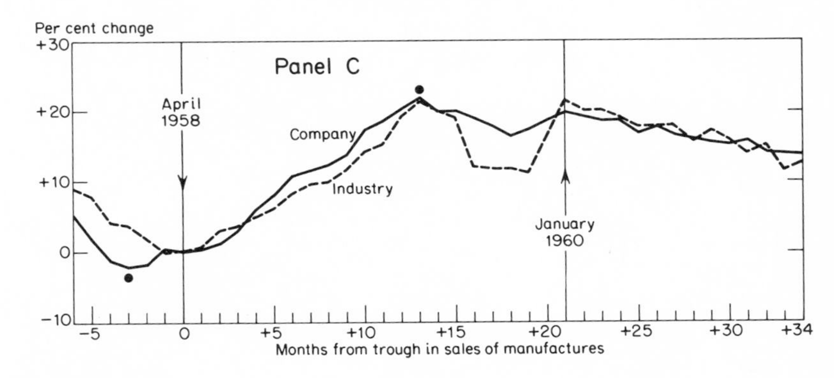

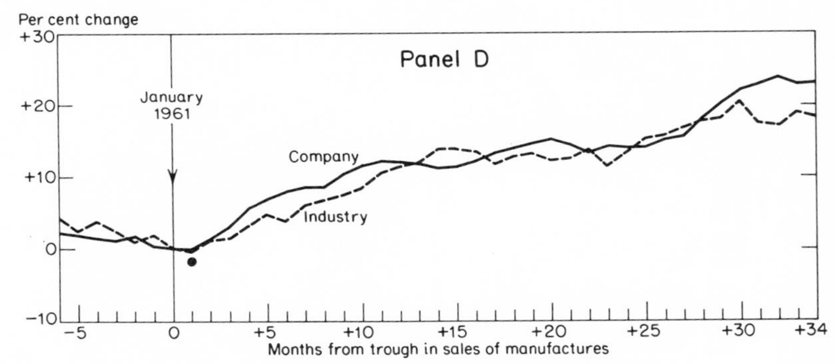

23 and Table 19 show that downswings in company sales tend to be shorter and milder than those of industry at large. In one case (1960–61), the company contraction appears to have lasted longer than the industry contraction. This is due to a double peak in company sales. If the later peak were recognized, timing and duration of company and industry sales would be virtually the same. The analyst of company sales will, of course, be interested in determining the conditions under which the company does better or worse than its industry, and he will attempt to utilize the resultant insights for forecasting and, perhaps, for suggestions to management.

Thus the described techniques of intercyclical comparisons of recessions and recoveries have wide applications. It is obvious that these applications are greatly facilitated by the availability of electronic

TABLE 18

COMPARATIVE CYCLICAL BEHAVIOR DURING FOUR RECOVERIES, SALES OF A MANUFACTURING COMPANY AND SALES OF ALL MANUFACTURES, 1949–65

Recovery Starting[a]	*Timing, Company Compared to Industry, Leads (−), Lags (+), Coincidences (0), in Months*		*Duration of Specific Expansion (months)*			*Amplitudes in Percentage of Trough Levels*[b]		
						Company		*Industry*
	At Initial Trough	*At Subsequent Peak*	*Company*	*Industry*	*Company minus Industry*	*Industry Dates*	*Company Dates*	*Industry Dates*
10/49	−5 months	−3 months	47	45	+2	+55	+70	+61
9/54	−8 months	+1 month	38	29	+9	+24	+35	+28
4/58	−3 months	−8 months[c]	16	21	−5	+19	+23	+20
1/61[d]	+1 month					+53	+52	+36

[a] These dates are troughs of sales in all manufacturing.

[b] Three-month average at peak minus three-month average at trough, as percentage of the former. Peak and trough dates are those of manufacturing sales for the first and the last column; they refer to turning points in company sales for the middle column.

[c] Double peak; use of the second high would lead to coincident timing.

[d] Measured to December 1965, last available date.

CHART 23

RECESSION ANALYSIS,
COMPANY SALES AND SALES OF MANUFACTURES,
PERCENTAGE CHANGE
FROM PEAK IN SALES OF MANUFACTURES, 1948–61

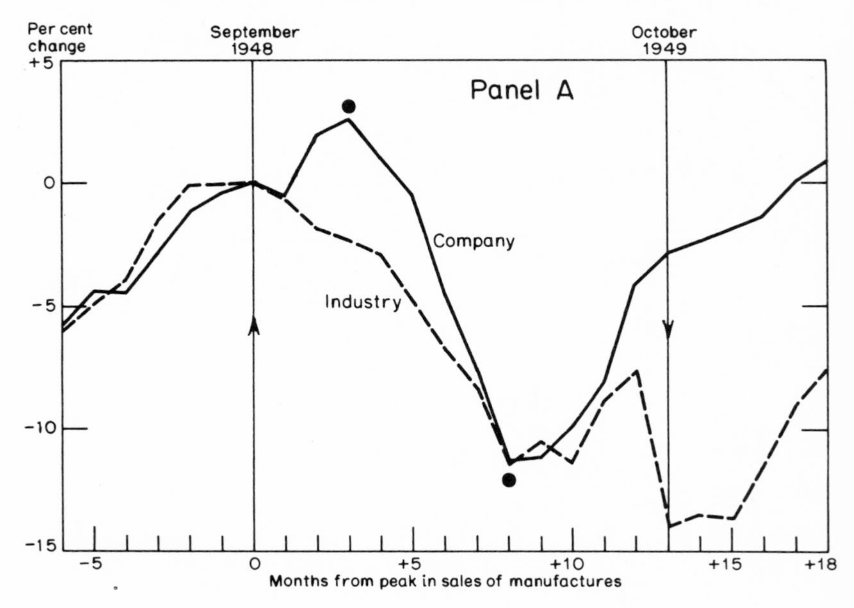

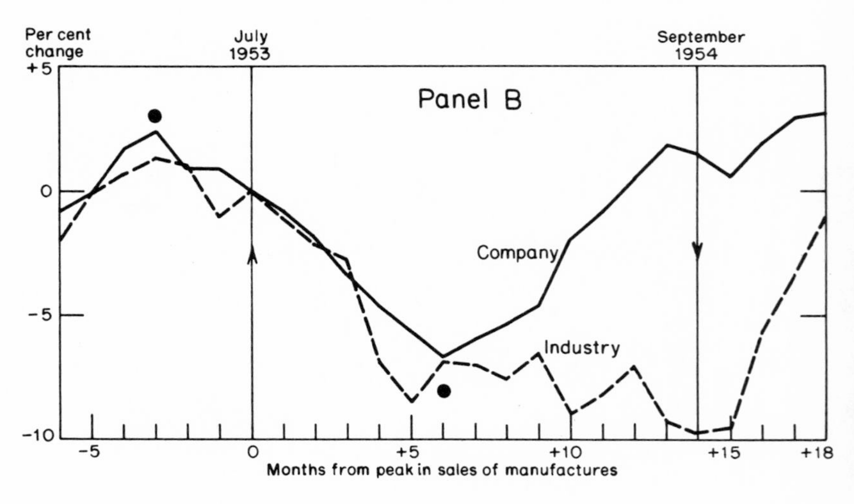

CHART 23

(Concluded)

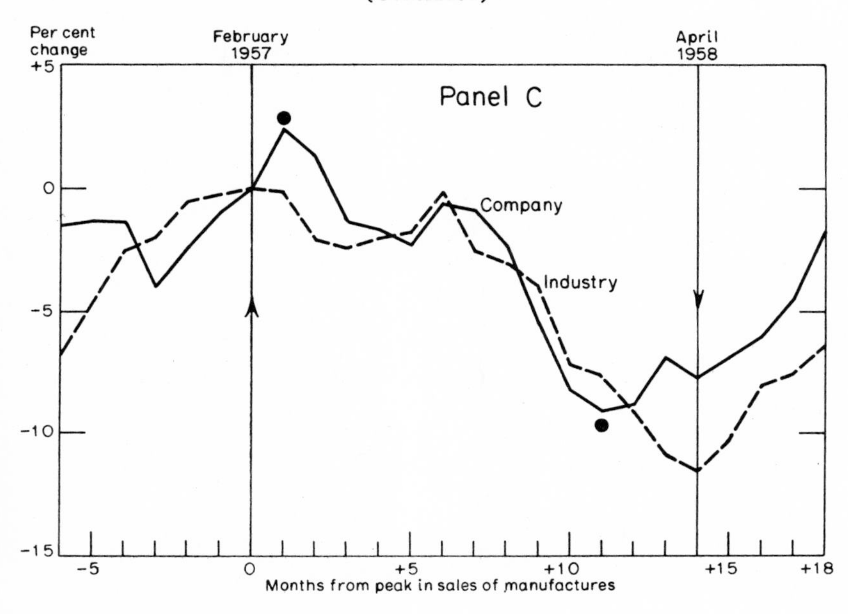

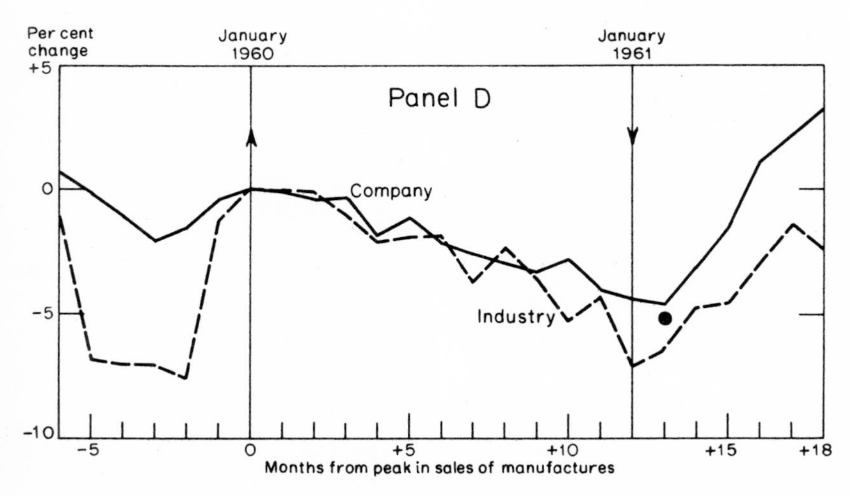

Note: Vertical lines denote months of cyclical turns in sales of manufactures; percentage changes are computed from three months averages centered around the trough (peak). Circles denote cyclical turning points in the company series.

TABLE 19

COMPARATIVE CYCLICAL BEHAVIOR DURING FOUR RECESSIONS, SALES OF A MANUFACTURING COMPANY AND SALES OF ALL MANUFACTURES, 1948–62

Recession Starting[a]	*Timing, Company Compared to Industry, Leads (−), Lags (+), Coincidences (0), in Months*		*Duration of Specific Contraction (months)*			*Amplitudes in Percentage of Peak Levels*[b]		
						Company		*Industry*
	At Initial Peak	*At Subsequent Trough*	*Company*	*Industry*	*Company minus Industry*	*Industry Dates*	*Company Dates*	*Industry Dates*
9/48	+3 months	−5 months	5	13	−8	−3	−10	−12
7/53	−3 months	−8 months	9	14	−5	+1	−8	−10
2/57	+1 month	−3 months	10	14	−4	−7	−9	−11
1/60	−8 months[c]	+1 month	21	12	+9	−4	−5	−6

[a] These dates are peaks of sales in all manufacturing.

[b] Three-month average at trough minus three-month average at peak, as percentage of the former. Peak and trough dates are those of manufacturing sales for the first and the last column; they refer to turning points in company sales for the middle column.

[c] Double peak; use of the second high would lead to coincident timing.

computer programs. Electronic processing becomes almost indispensable if the analysis is carried through with several variants used in conjunction with each other.

VARIANTS OF ANALYSIS

In the previous discussion, percentage changes were computed for increasing spans, measured from business cycle turns. This means that a common reference cycle framework was used for the analysis of all individual activities. For certain purposes it is preferable to compute cyclical changes from turning points of the series themselves, i.e., from their specific turns. Broadly speaking, the reference cycle version of recession-recovery analysis is preferable when the interest centers on the contribution of various activities to cyclical changes in business conditions at large, or when comparisons among a variety of activities are facilitated by analyzing each of them in a common framework. The specific cycle version provides a more relevant and more fruitful focus if interest is centered on the cyclical characteristics, sensitivity, or prospects of an individual activity—be it the fortunes of an industry, the profits of a company, or the sales of a product. In many cases, both types of analysis may be of interest; some measures resulting from the two types of analysis will be compared below.

Since employment is a well-conforming series and shows only short leads and lags relative to business cycle turns,[17] the difference between the reference version and the specific version of the analysis is not substantial. However, it may be very marked if the analysis is performed on series with long and irregular leads or lags. The difference between the two versions is illustrated in Chart 24, which shows comparative recovery behavior of new orders for durable goods during business cycle and specific cycle expansions. At first glance the two panels seem to have little in common. In the specific cycle version (lower panel) the recovery patterns vary widely. The recovery from the 1961 trough proceeds vigorously for about a year, stalls during the second year, and is resumed thereafter; the recovery after the 1953 trough, by contrast, does not really get under way for a year, and then shows good vigor for the rest of the reported period; in the recovery after 1958, an initial hesitance, very fast progress to the sixteenth month, and marked decline thereafter can be observed. By

[17] See pp. 80 ff.

CHART 24
RECOVERY ANALYSIS, NEW ORDERS FOR DURABLE GOODS, PERCENTAGE CHANGE FROM PRECEDING REFERENCE AND SPECIFIC CYCLE PEAK, 1954–64

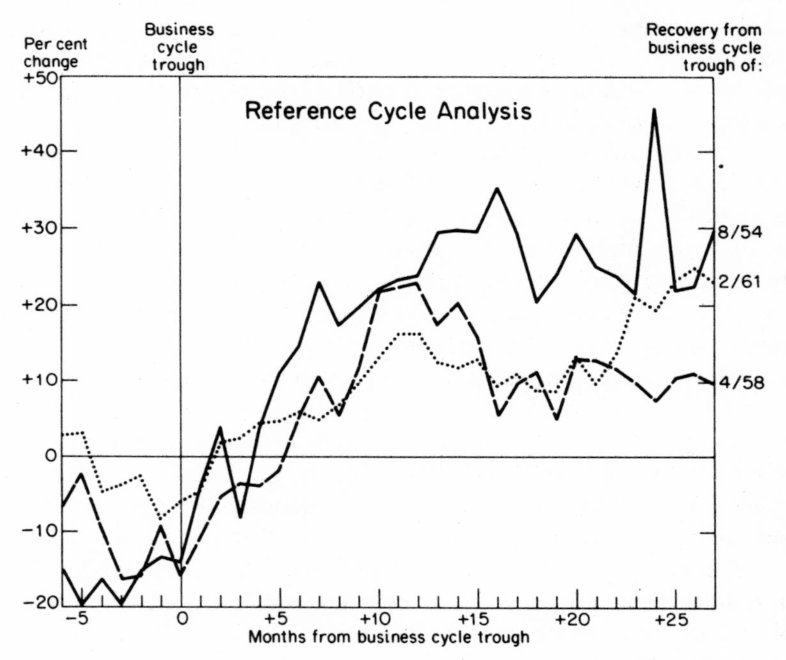

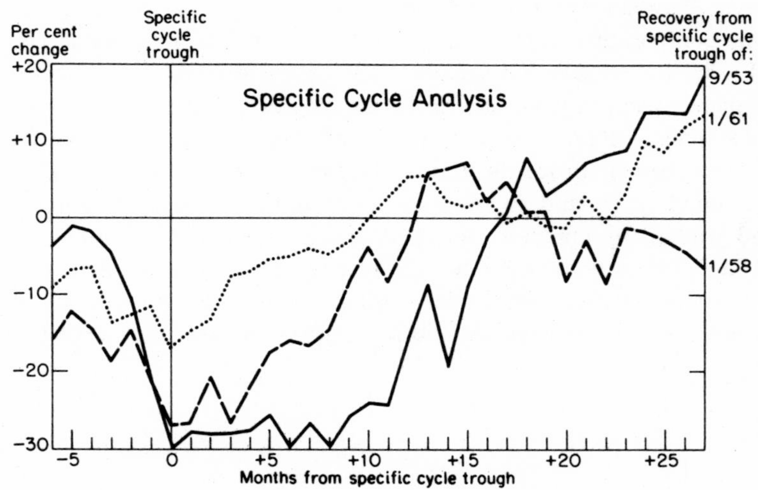

contrast, the reference cycle analysis (upper panel) shows rather similar recovery movements of new orders for about a year and a tendency toward reversals for the better part of the next year. Only thereafter can strikingly different developments be seen. A close look at the comparative behavior of new orders during the expansion starting about 1954 will be helpful. In the upper panel (which shows recovery from business cycle troughs, relative to preceding business cycle peak levels), new orders recover early and fast, reach previous peak levels soon,[18] and continue to rise at a more rapid rate than that shown during the other two recoveries. By contrast, in the lower panel (which shows recovery from the lowest level of new orders themselves, relative to their own preceding peak levels) the comparative performance looks quite different. New orders, after their own low, scarcely recover at all during the first eight months; they reach and exceed previous peak levels only after the sixteenth month. At the twenty-eighth month the expansion is still in full swing, while the other two depicted expansions lasted thirteen and fifteen months. Which presentation tells the true story about comparative performance? Obviously neither. Both show different aspects of cyclical behavior, and the very difference of the patterns demonstrates how unsatisfactory it might be to base one's evaluation on only one of the versions.

Let us pursue the comparison a bit further. The two representations vary in only two respects—the base for the computation of relatives and the way in which the series are chronologically aligned. The differences of the percentage bases (levels at business cycle peaks compared with those at specific cycle peaks) are usually not very large, even if there are relatively long leads or lags. This is due to the base being at similar long-term and, usually, at roughly similar cyclical levels. In the example, new orders at the 1953 reference peak amount to $12.1 billion, at the corresponding specific peak to $13.8 billion, about 14 per cent higher. The apparent performance is more importantly affected by the change in chronological alignment. In the reference cycle version, the analysis aligns the series so that the months of the reference turns (peaks for recession and troughs for recovery analysis) are at the origin of the horizontal scale; in the specific cycle analysis, the same holds for the cyclical turns of the series itself. This may lead to substantial differences in alignment between the two

[18] This occurs when the deviations from previous peak levels become zero.

analyses. New orders experienced their own trough in September 1953, eleven months before the business cycle trough in August 1954. This long lead, together with less drastic leads at the other troughs, accounts for the strong differences in comparative recovery patterns. This situation highlights a problem that has been discussed earlier in a different context,[19] that is, the sensitivity of cyclical analysis to the determination of turning points—particularly when flat-bottom troughs of substantial duration are experienced. The second panel of Chart 24 shows that new orders maintained a low level from September 1953 (0 on the scale) to May 1954 (+8 on the scale). There is no doubt that the September 1953 level was lower and correctly chosen as the specific trough. But how significant was the difference, in view of the considerable random fluctuations exhibited by the series? The choice of May 1954 as the specific trough would have substantially affected the analysis of comparative behavior. The specific recovery would have been very much more favorable, compared to the later ones. The moral of this discussion is, of course, that recession-recovery analysis —as any other analytical tool—must not be used mechanically. The electronic computer output provides sufficient information for recognizing the effects of marginal decisions and (if necessary) for evaluating the effects of alternatives. It is very inexpensive to run recession-recovery analysis for alternative sets of chronologies once the basic input has been prepared.

Although for many purposes the comparative analysis is best made in the form of percentage changes or relatives, this is not necessarily always the case. In certain circumstances, the changes may be computed and compared in "absolute" form, that is, in terms of the units in which the original values are stated. The reasons for preferring the absolute form are varied. One is purely technical: If the series contains negative numbers (as is likely, for instance, in a series of budget surpluses and deficits or a series on inventory change), percentage changes cannot be computed, or they may become awkwardly large. Also, when the units have independent meaning and are easily evaluated against a standard (length of the average workweek), absolute changes may be desirable. For series that are components of a total (such as the components of GNP), comparisons of the absolute changes may be of interest. Finally, if the original units are already in ratio

[19] See pp. 12 ff.

form (as in the case of unemployment rates, capacity utilization rates, or interest rates), the percentage-change analysis may be less instructive than the analysis in terms of the rates themselves. Here, both the absence of strong trends and the presence of strong benchmark standards make changes in the absolute units directly comparable. Except for series containing negative numbers, the case for absolute differences is not really hard and fast. There is, of course, always the possibility of performing the analysis in both ways, a possibility that has become more attractive with the availability of an electronic computer program for both versions of the analysis.

On occasion, one may wish to perform the comparative analysis of cyclical behavior on the basis of levels rather than changes. Chart 25 illustrates this version for the unemployment rate during recent expansions. The vertical scale shows unemployment as a percentage of the labor force in the original units of the analyzed series (in contrast to Chart 18, where this scale shows percentage deviations from previous peak levels). This presentation is distinguished from that of a conventional time series merely by alignment around business cycle troughs. Note that the unemployment rate, after about two years of recovery in general business conditions, showed a historical sequence of increasingly higher levels. This is the inverse order of the comparable lines on Chart 20, which showed the degree to which previous prosperity levels were approximated. Again the computer program, which provides all versions of recession-recovery analysis, permits a view of many aspects of cyclical behavior.

Comparative analysis by the described techniques is easily impaired if the original series exhibit strong irregular movements. This tends to occur when sensitive activities (such as construction contracts, new orders, business failures, and similar indicators) are the raw material for the analysis. It may be a still more serious problem if the analysis is applied to data of rather narrow coverage, such as industry or company information. A first step in reducing the undesired preponderance of the irregular element is to compute changes only for selected spans (three, six, nine, or four, eight, twelve months, etc.) so that the cyclical forces have a chance to assert themselves, over given intervals, against the irregular ones. This is, however, more an expository than an analytical device. It reduces confusion but does not reduce the absolute size of the random component of the observation, and for current analysis it prevents use of the most recent information. In

CHART 25

RECOVERY ANALYSIS, UNEMPLOYMENT RATE, ABSOLUTE LEVELS, ARRANGED AROUND BUSINESS CYCLE TROUGHS, 1949–63

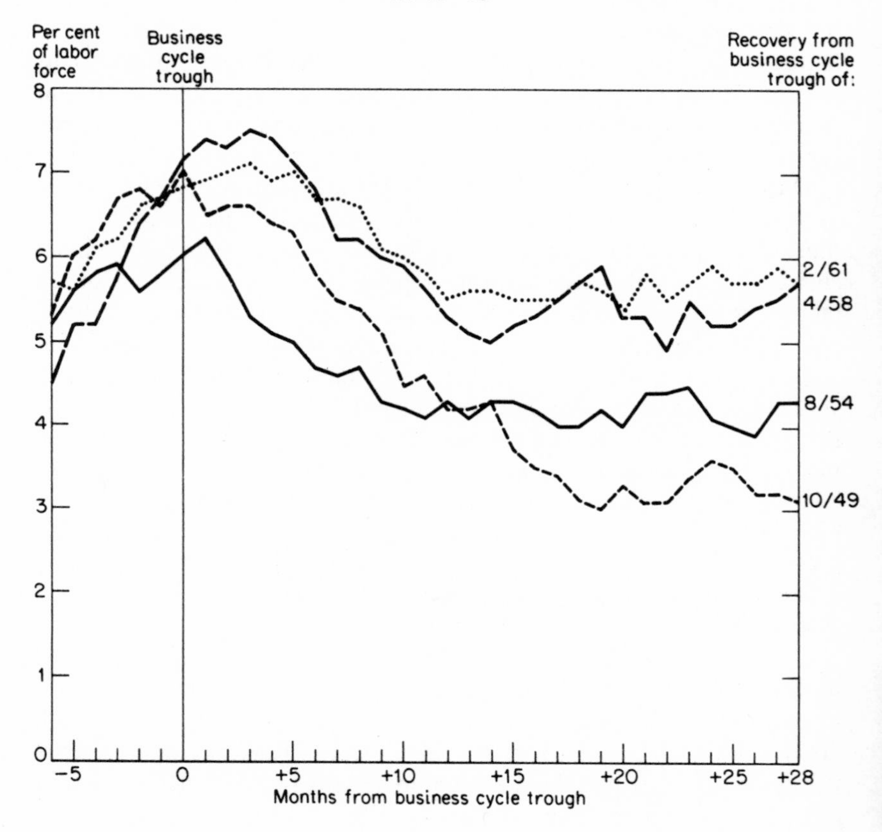

many cases, the use of a short smoothing term[20] provides an answer, and the resultant loss of "currency" of the analysis may be a small cost compared to the greater cyclical significance of the computed measures. Also, the unsmoothed data can always be used side by side with the smoothed data.

[20] Some programmed seasonal adjustment procedures provide a smoothed version of the adjusted series.

INTERPRETATION OF OUTPUT

The major goal of this part of the study is to offer guidance for the understanding and use of the programmed versions of two closely related analytical techniques. An exposition of the general approach and the major versions of recession-recovery analysis has been provided; now it is time to turn to the programmed output. The output tables for recession analysis are found in Appendix 4A, for recovery analysis in Appendix 4B. Both appendixes contain tables pertaining to reference analysis (R) and specific analysis (S). The designation of output tables specifies the appendix number, the table number, and the type of analysis (R or S).

The first table of Appendix 4A (Output Table 4A-1-R) presents the time series being analyzed in original units.[21]

Output Table 4A-2-R contains reference phase amplitudes in absolute form, which in the present case means changes in employment during reference contractions and expansions. Output Table 4A-3-R contains analogous information, in terms of percentage changes relative to previous turns. The importance of these tables is that they permit investigation of the degree to which the incomplete amplitudes of recessions, i.e., the amplitudes for specified chronological portions of recessions, reported in the main table of the analysis (Output Table 4A-5-R on absolute and 4A-6-R on relative changes), are associated with the full amplitudes reported here.

Readers of the first part of this book, on the standard business cycle analysis, must be cautioned not to mistake the amplitude measures used here for those of the standard analysis. The phase amplitudes of recession analysis are conventional percentage changes from the three-month average centered on the peak month to the corresponding average at the subsequent trough. These amplitudes will usually deviate a bit from the relative change from peak standing to trough month, as reported in later tables. The reason is that the later tables show the decline to the trough *month* rather than to the three-month average centered at the trough. Observe that the reference amplitude of employment during the Great Depression, 1929–33, is −30.7 per cent

[21] The table presents the data input for analysis. A seasonally adjusted series is used if such adjustment was deemed necessary. Otherwise the unadjusted series, or possibly a smoothed version of the adjusted or unadjusted series, is used and printed.

according to Output Table 4A-3-R, and the corresponding decline from the 1929 peak to March 1933 amounts to 31.3 per cent according to Output Table 4A-6-R.[22] The latter measure is comparable to the measures for other spans. The usefulness of the phase amplitudes for comparison with the cyclical declines of shorter duration will be demonstrated later on.

The next table (Output Table 4A-4-R) contains some of the same data as Output Table 4A-1-R, but they are ordered according to their chronological relation to the relevant cyclical turns—which, in the present case, are business cycle peaks (reference cycle peaks). The data are still in original units. The first panel contains the data for the eleven months preceding each peak.[23] The provision of data for some period before peaks is particularly valuable in the case of reference cycle analysis because it permits the user to be informed about the behavior of a series in the entire neighborhood of a turn in general business conditions. In the case of specific analysis, it would be known at least that the movement before the peak was upward (or sideways), which of course is not very precise information. In the case of reference analysis, nothing would be known about the movement before the turn, in the absence of prior data. The data for eleven months before the turn may frequently include the upper turn of leading series but not, of course, the turns of series with leads of one year or more.

The programmed recession analysis can provide data for up to five years after each cyclical peak. This may seem inordinately long, in view of the fact that the duration of reference recessions since World War I averages only about fifteen months. However, the 1929–33 recession lasted forty-three months; lagging series may continue to decline for many months after business cycle troughs; and the durations of contractions in specific activities can markedly exceed those in general business conditions.[24] However, since cycle phases are often

[22] The reference trough, forty-three months after the reference peak, is marked by an asterisk. The given value, 68.7, is 31.3 per cent below the preceding peak level.

[23] A period of eleven, rather than the more plausible twelve, months before the turn was selected for technical convenience, i.e., to accommodate the value for the turn itself (the zero month) on the same table.

[24] Changes in consumer instalment debt, for example, declined for fully three years between 1955 and 1958, a period four times as long as the associated business cycle decline from July 1957 to April 1958. Also, for series that bear an inverted relation to business activity (such as the unemployment rate), the specific cycle declines correspond to business expansions and are frequently quite long.

shorter, the user can decide whether he wants information for more than two years. Of course, as the interval becomes longer, the meaningfulness of the comparison in terms of its relation to the starting date becomes more dubious.

The next tables, Output Tables 4A-5-R and 4A-6-R, present the key measures of recession analysis, the changes from peak levels. These changes may be in absolute form (4A-5-R) or in terms of percentages (4A-6-R).[25]

Some entries on Output Tables 4A-5-R and 4A-6-R have asterisks. These asterisks identify the values at the cyclical turn following the initial turn (upon which the analysis is based). In the example, the asterisks show the values at the reference troughs following the reference peaks from which the declines were measured. The identification is valuable because it helps to delineate periods that may be important for the comparisons intended—in the present case the asterisks denote the termination of the reference contractions.

Each panel of these two tables ends with some lines giving totals, averages, and average deviations. These measures refer to the changes from reference peaks, as printed in the panel. The totals are merely intermediate computational results which are printed out to facilitate modification of the averages (for instance, by excluding cycles). The averages are unweighted means of the previously reported changes. They also contain, of course, rises (or relatives above 100) which become increasingly frequent after some of the contraction phases have reached their end. This averaging process does not stop after the duration of the shortest business cycle contraction, so that, after a while, the reported averages include more and more experience reaching into cyclical expansions. Consequently, the behavior of these averages must be interpreted with great care. Reference to the asterisks in the body of the percentage change table will help in this interpretation. The last line of each change panel provides average deviations of the changes (or relatives) from the reported means. These average deviations are small in the neighborhood of the initial peak and tend to increase as the differential paths of the various recessions become more pronounced. They serve in the evaluation of the representativeness of the averages and might also help in comparing the efficiency

[25] Output Table 4A-6-R is expressed in terms of relatives, with the peak standing as base. The difference is, of course, purely formal, since percentage changes are simply relatives minus 100.

of the standard business cycle analysis with that of recession-recovery analysis for the purpose of describing typical behavior.

Appendix 4A also contains the specific cycle version of recession analysis. The output tables are numbered from 4A-1-S to 4A-6-S. Since the output is presented in the same format as that for the reference cycle version, comments will be brief. The absolute and percentage changes (or relatives) measure the changes from the specific peak relative to the levels at that peak, for increasing spans. The asterisks in Output Tables 4A-5-S and 4A-6-S refer to the subsequent specific troughs. These tables show that before the specific-peak date, the reported standings were generally below the peak standings—a fact that distinguishes these tables from Output Table 4A-4-R, where, during some of the shorter spans before the reference peak, the standings were above peak standings.[26] Otherwise, comparison between the two tables reveals resemblances rather than differences—due largely to the close timing relationship between cyclical peaks in nonagricultural employment and cyclical peaks in general business activity.

Appendix 4B refers to recovery analysis. Tables 4B-1-R to 4B-6-R for reference analysis correspond to the similarly numbered tables of the recession analysis. The analogous Tables 4B-1-S to 4B-6-S for specific analysis are not included in the Appendix. Changes are computed from initial troughs, reference or specific. Appendix 4B also contains Output Tables 4B-7 and 4B-8, showing changes computed in terms of the levels of the preceding peak, reference (R) or specific (S). In all output tables for recovery analysis, provision is made for comparisons up to five years after the reference or specific trough dates. This extends, of course, considerably beyond the recovery period proper, as previously defined, and is actually designed to permit comparison throughout most complete expansion periods. Most expansions in the United States have lasted less than five years. Since many expansions had considerably shorter durations, the percentage changes extending into the subsequent contraction phase may be without interest. Asterisks show the end of the expansion phases. The user may disregard the later data. He can also specify the number of years, after troughs, to be covered by the analysis.

[26] Note that in the change tables, relatives below 100 *before* the peak denote *rises to* the peak; *after* the peak, they denote *declines from* the peak. Analogously, negative entries before the peak denote rises: after the peak they denote declines.

POSTSCRIPT

During the latter part of the 1960's, recession-recovery analysis fell into disuse. The reasons are simple enough. After the 1961 trough there was no decline in the economy prompting the questions which recession analysis is designed to answer. And the long boom soon outlasted other postwar expansions with which meaningful comparisons could be made. Thus, for recession analysis there was no recession to be dealt with and for recovery analysis there were no recoveries to be compared to. All this changed with the onset of the slowdown which started late in 1969. Some of the classical questions were asked in new form: Is there going to be a recession—mini, midi, or perhaps even maxi? Whatever it is, how long will it last? Was there an upper turning point, and if so, when did it occur? Recession analysis came into vogue again, at least as a tool of observation.

Doubts about recession analysis as a forecasting tool arose, since the future of this man-made slowdown seemed to be so highly dependent on the mix of monetary, fiscal, and other policies. However, it turned out that neither the time lags nor the magnitudes of the effects of these policies can be foretold with any precision. Furthermore, economic fortunes are substantially modified by private sector decisions, which are only loosely related to the federal policies. Thus, the questions of how the economy would behave were wide open and, in fact, subject to spirited debate. These circumstances brought recession analysis into use again, and the requests for data, programs, and procedural guidance mounted.

At the time of this writing—September 1970—it looks as if recovery analysis may soon become the appropriate tool. In spite of our proprietary interest in its application, we would not mind if it again lost its applicability by our running out of expansions with comparable durations.

APPENDIX TO CHAPTER 4

A

SAMPLE RUN, RECOVERY ANALYSIS

Output Table 4A-1R

NBER RECESSION-RECOVERY ANALYSIS

REFERENCE ANALYSIS

EMPLOYEES IN NONAG ESTABLISHMENTS

HUNDRED THOUSAND PERSONS 8268

YEAR	JAN	FEB	MAR	APR	MAY	JUNE	JULY	AUG	SEPT	OCT	NOV	DEC
1929	326.0	326.0	328.0	329.0	330.0	331.0	332.0	333.0	331.0	330.0	327.0	323.
1930	319.0	317.0	314.0	313.0	311.0	308.0	304.0	300.0	297.0	294.0	291.0	289.
1931	286.0	285.0	283.0	282.0	280.0	277.0	275.0	272.0	268.0	265.0	262.0	260.
1962	547.0	550.0	552.0	554.0	555.0	556.0	557.0	557.0	558.0	558.0	559.0	559.
1963	559.0	560.0	562.0	564.0	565.0	566.0	568.0	568.0	569.0	571.0	571.0	573.

Output Table 4A-2R

ABSOLUTE CHANGES BETWEEN CYCLICAL TURNS

PEAK	TROUGH	PEAK	PEAK	TROUGH	PEAK	FALL	RISE
1929 8	1933 3	1937 5	332.0	230.0	319.3	-102.0	89.3
1937 5	1938 6	1945 2	319.3	287.3	417.7	-32.0	130.3
1945 2	1945 10	1948 11	417.7	385.7	451.0	-32.0	65.3
1948 11	1949 10	1953 7	451.0	432.3	503.7	-18.7	71.3
1953 7	1954 8	1957 7	503.7	487.0	530.0	-16.7	43.0
1957 7	1958 4		530.0	509.7		-20.3	
TOTAL						-221.7	399.3
AVERAGE						-36.9	79.9
AVERAGE DEVIATIONS						21.7	24.0

OMIT THE FOLLOWING CYCLES

1945 2 1945 10 1948 11

	FALL	RISE
TOTAL	-189.7	334.0
AVERAGE	-31.6	66.8
AVERAGE DEVIATIONS	18.3	22.9

Output Table 4A-3R

RELATIVE CHANGES BETWEEN CYCLICAL TURNS

PEAK	TROUGH	PEAK	PEAK	TROUGH	PEAK	FALL	RISE
1929 8	1933 3	1937 5	332.0	230.0	319.3	-30.7	38.8
1937 5	1938 6	1945 2	319.3	287.3	417.7	-10.0	45.4
1945 2	1945 10	1948 11	417.7	385.7	451.0	-7.7	16.9
1948 11	1949 10	1953 7	451.0	432.3	503.7	-4.1	16.5
1953 7	1954 8	1957 7	503.7	487.0	530.0	-3.3	8.8
1957 7	1958 4		530.0	509.7		-3.8	
TOTAL						-59.7	126.5
AVERAGE						-9.9	25.3
AVERAGE DEVIATIONS						6.9	13.4

OMIT THE FOLLOWING CYCLES

1945 2 1945 10 1948 11

	FALL	RISE
TOTAL	-52.0	109.5
AVERAGE	-8.7	21.9
AVERAGE DEVIATIONS	6.4	11.8

Output Table 4A-4R

EMPLOYEES IN NONAG ESTABLISHMENTS

HUNDRED THOUSAND PERSONS 8268

REFERENCE ANALYSIS

STANDINGS

DATE OF EAK	STANDING AT PEAK	-11 MO	-10 MO	STANDING -9 MO	-8 MO	-7 MO	-6 MO	ONE YEAR BEFORE PEAK -5 MO	-4 MO	-3 MO	-2 MO	-1 MO	0 MO
929 8	332.00	0.	0.	0.	0.	326.	326.	328.	329.	330.	331.	332.	333.
937 5	319.33	298.	301.	303.	305.	306.	309.	312.	313.	315.	317.	318.	320.
945 2	417.67	422.	419.	418.	417.	417.	417.	416.	416.	416.	417.	418.	418.
948 11	451.00	446.	447.	445.	447.	443.	447.	449.	451.	450.	451.	451.	451.
953 7	503.67	487.	491.	495.	497.	500.	501.	503.	504.	504.	504.	504.	504.
957 7	530.00	525.	524.	527.	528.	529.	529.	531.	531.	531.	530.	530.	530.

Output Table 4A-5R

DATE OF EAK	STANDING AT PEAK	-11 MO	-10 MO	ABSOLUTE CHANGE -9 MO	-8 MO	-7 MO	-6 MO	ONE YEAR BEFORE PEAK -5 MO	-4 MO	-3 MO	-2 MO	-1 MO	0 MO
929 8	332.00	0.0	0.0	0.0	0.0	-6.0	-6.0	-4.0	-3.0	-2.0	-1.0	0.0	1.0
937 5	319.33	-21.3	-18.3	-16.3	-14.3	-13.3	-10.3	-7.3	-6.3	-4.3	-2.3	-1.3	0.7
945 2	417.67	4.3	1.3	0.3	-0.7	-0.7	-0.7	-1.7	-1.7	-1.7	-0.7	0.3	0.3
948 11	451.00	-5.0	-4.0	-6.0	-4.0	-8.0	-4.0	-2.0	0.0	-1.0	0.0	0.0	0.0
953 7	503.67	-16.7	-12.7	-8.7	-6.7	-3.7	-2.7	-0.7	0.3	0.3	0.3	0.3	0.3
57 7	530.00	-5.0	-6.0	-3.0	-2.0	-1.0	-1.0	1.0	1.0	1.0	0.0	0.0	0.0
OTAL		-43.7	-39.7	-33.7	-27.7	-32.7	-24.7	-14.7	-9.7	-7.7	-3.7	-0.7	2.3
VERAGE		-8.7	-7.9	-6.7	-5.5	-5.4	-4.1	-2.4	-1.6	-1.3	-0.6	-0.1	0.4
VE DEVIATION		8.2	6.1	4.6	4.0	3.7	2.7	2.1	2.1	1.4	0.7	0.4	0.3

OMIT THE FOLLOWING CYCLES

1945 2 1945 10 1948 11

	-11 MO	-10 MO	-9 MO	-8 MO	-7 MO	-6 MO	-5 MO	-4 MO	-3 MO	-2 MO	-1 MO	0 MO
OTAL	-48.0	-41.0	-34.0	-27.0	-32.0	-24.0	-13.0	-8.0	-6.0	-3.0	-1.0	2.0
VERAGE	-12.0	-10.2	-8.5	-6.7	-6.4	-4.8	-2.6	-1.6	-1.2	-0.6	-0.2	0.4
VE DEVIATION	7.0	5.3	4.0	3.8	3.4	2.7	2.5	2.5	1.6	0.9	0.5	0.3

Output Table 4A-6R

ATE OF AK	STANDING AT PEAK	-11 MO	-10 MO	RELATIVE STANDING -9 MO	-8 MO	-7 MO	-6 MO	ONE YEAR BEFORE PEAK -5 MO	-4 MO	-3 MO	-2 MO	-1 MO	0 MO
29 8	332.00	0.0	0.0	0.0	0.0	98.2	98.2	98.8	99.1	99.4	99.7	100.0	100.3
37 5	319.33	93.3	94.3	94.9	95.5	95.8	96.8	97.7	98.0	98.6	99.3	99.6	100.2
45 2	417.67	101.0	100.3	100.1	99.8	99.8	99.8	99.6	99.6	99.6	99.8	100.1	100.1
48 11	451.00	98.9	99.1	98.7	99.1	98.2	99.1	99.6	100.0	99.8	100.0	100.0	100.0
53 7	503.67	96.7	97.5	98.3	98.7	99.3	99.5	99.9	100.1	100.1	100.1	100.1	100.1
57 7	530.00	99.1	98.9	99.4	99.6	99.8	99.8	100.2	100.2	100.2	100.0	100.0	100.0
TAL		489.0	490.0	491.3	492.8	591.2	593.2	595.7	597.0	597.7	598.9	599.7	600.7
ERAGE		97.8	98.0	98.3	98.6	98.5	98.9	99.3	99.5	99.6	99.8	100.0	100.1
E DEVIATION		2.2	1.7	1.4	1.2	1.1	0.9	0.7	0.6	0.4	0.2	0.1	0.1

OMIT THE FOLLOWING CYCLES

1945 2 1945 10 1948 11

	-11 MO	-10 MO	-9 MO	-8 MO	-7 MO	-6 MO	-5 MO	-4 MO	-3 MO	-2 MO	-1 MO	0 MO
TAL	388.0	389.7	391.3	392.9	491.3	493.4	496.1	497.4	498.1	499.0	499.6	500.6
ERAGE	97.0	97.4	97.8	98.2	98.3	98.7	99.2	99.5	99.6	99.8	99.9	100.1
E DEVIATION	2.0	1.6	1.5	1.4	1.0	1.0	0.8	0.7	0.5	0.3	0.1	0.1

Output Table 4A-4R

EMPLOYEES IN NONAG ESTABLISHMENTS

HUNDRED THOUSAND PERSONS 8268

REFERENCE ANALYSIS

DATE OF	STANDING			STANDINGS STANDING				FIRST YEAR AFTER PEAK					
PEAK	AT PEAK	+1 MO	+2 MO	+3 MO	+4 MO	+5 MO	+6 MO	+7 MO	+8 MO	+9 MO	+10 MO	+11 MO	+12 MO
1929 8	332.00	331.	330.	327.	323.	319.	317.	314.	313.	311.	308.	304.	300 [illegible]
1937 5	319.33	320.	321.	321.	320.	317.	312.	305.	299.	296.	294.	292.	288 [illegible]
1945 2	417.67	417.	413.	411.	409.	406.	403.	384.	385.	388.	390.	397.	392 [illegible]
1948 11	451.00	451.	446.	444.	442.	441.	438.	436.	435.	435.	437.	428.	432 [illegible]
1953 7	503.67	503.	502.	501.	498.	497.	494.	493.	491.	490.	489.	488.	48 [illegible]
1957 7	530.00	530.	528.	527.	525.	523.	521.	515.	512.	509.	508.	509.	50 [illegible]

Output Table 4A-5R

DATE OF	STANDING			ABSOLUTE CHANGE				FIRST YEAR AFTER PEAK					
PEAK	AT PEAK	+1 MO	+2 MO	+3 MO	+4 MO	+5 MO	+6 MO	+7 MO	+8 MO	+9 MO	+10 MO	+11 MO	+12 [illegible]
1929 8	332.00	-1.0	-2.0	-5.0	-9.0	-13.0	-15.0	-18.0	-19.0	-21.0	-24.0	-28.0	-32 [illegible]
1937 5	319.33	0.7	1.7	1.7	0.7	-2.3	-7.3	-14.3	-20.3	-23.3	-25.3	-27.3	-31 [illegible]
1945 2	417.67	-0.7	-4.7	-6.7	-8.7	-11.7	-14.7	-33.7	-32.7*	-29.7	-27.7	-20.7	-25 [illegible]
1948 11	451.00	0.0	-5.0	-7.0	-9.0	-10.0	-13.0	-15.0	-16.0	-16.0	-14.0	-23.0*	-19 [illegible]
1953 7	503.67	-0.7	-1.7	-2.7	-5.7	-6.7	-9.7	-10.7	-12.7	-13.7	-14.7	-15.7	-16 [illegible]
1957 7	530.00	0.0	-2.0	-3.0	-5.0	-7.0	-9.0	-15.0	-18.0	-21.0*	-22.0	-21.0	-21 [illegible]
TOTAL		-1.7	-13.7	-22.7	-36.7	-50.7	-68.7	-106.7	-118.7	-124.7	-127.7	-135.7	-145 [illegible]
AVERAGE		-0.3	-2.3	-3.8	-6.1	-8.4	-11.4	-17.8	-19.8	-20.8	-21.3	-22.6	-24 [illegible]
AVE DEVIATION		0.5	1.7	2.4	2.8	3.1	2.8	5.4	4.5	4.0	4.6	3.5	5 [illegible]

OMIT THE FOLLOWING CYCLES

1945 2 1945 10 1948 11

	+1 MO	+2 MO	+3 MO	+4 MO	+5 MO	+6 MO	+7 MO	+8 MO	+9 MO	+10 MO	+11 MO	+12 MO
TOTAL	-1.0	-9.0	-16.0	-28.0	-39.0	-54.0	-73.0	-86.0	-95.0	-100.0	-115.0	-12 [illegible]
AVERAGE	-0.2	-1.8	-3.2	-5.6	-7.8	-10.8	-14.6	-17.2	-19.0	-20.0	-23.0	-2 [illegible]
AVE DEVIATION	0.5	1.4	2.2	2.7	3.0	2.6	1.7	2.3	3.3	4.5	3.7	[illegible]

Output Table 4A-6R

DATE OF	STANDING			RELATIVE STANDING				FIRST YEAR AFTER PEAK					
PEAK	AT PEAK	+1 MO	+2 MO	+3 MO	+4 MO	+5 MO	+6 MO	+7 MO	+8 MO	+9 MO	+10 MO	+11 MO	+12
1929 8	332.00	99.7	99.4	98.5	97.3	96.1	95.5	94.6	94.3	93.7	92.8	91.6	9 [illegible]
1937 5	319.33	100.2	100.5	100.5	100.2	99.3	97.7	95.5	93.6	92.7	92.1	91.4	9 [illegible]
1945 2	417.67	99.8	98.9	98.4	97.9	97.2	96.5	91.9	92.2*	92.9	93.4	95.1	9 [illegible]
1948 11	451.00	100.0	98.9	98.4	98.0	97.8	97.1	96.7	96.5	96.5	96.9	94.9*	9 [illegible]
1953 7	503.67	99.9	99.7	99.5	98.9	98.7	98.1	97.9	97.5	97.3	97.1	96.9	9 [illegible]
1957 7	530.00	100.0	99.6	99.4	99.1	98.7	98.3	97.2	96.6	96.0*	95.8	96.0	9 [illegible]
TOTAL		599.6	597.0	594.8	591.4	587.7	583.2	573.8	570.6	569.0	568.0	565.9	56 [illegible]
AVERAGE		99.9	99.5	99.1	98.6	97.9	97.2	95.6	95.1	94.8	94.7	94.3	9 [illegible]
AVE DEVIATION		0.1	0.4	0.7	0.8	0.9	0.8	1.6	1.7	1.8	1.9	1.9	[illegible]

OMIT THE FOLLOWING CYCLES

1945 2 1945 10 1948 11

	+1 MO	+2 MO	+3 MO	+4 MO	+5 MO	+6 MO	+7 MO	+8 MO	+9 MO	+10 MO	+11 MO	+12 MO
TOTAL	499.8	498.1	496.4	493.4	490.5	486.7	481.8	478.5	476.1	474.7	470.8	4 [illegible]
AVERAGE	100.0	99.6	99.3	98.7	98.1	97.3	96.4	95.7	95.2	94.9	94.2	[illegible]
AVE DEVIATION	0.1	0.4	0.6	0.8	0.9	0.8	1.1	1.4	1.6	2.0	2.1	[illegible]

Output Table 4A-4S

EMPLOYEES IN NONAG ESTABLISHMENTS

HUNDRED THOUSAND PERSONS 8268

SPECIFIC ANALYSIS

DATE OF PEAK	STANDING AT PEAK	-11 MO	-10 MO	STANDINGS STANDING -9 MO	-8 MO	-7 MO	-6 MO	ONE YEAR BEFORE PEAK -5 MO	-4 MO	-3 MO	-2 MO	-1 MO	0 MO
929 8	332.00	0.	0.	0.	0.	326.	326.	328.	329.	330.	331.	332.	333.
937 7	320.67	303.	305.	306.	309.	312.	313.	315.	317.	318.	320.	320.	321.
943 11	426.33	418.	421.	423.	424.	424.	424.	426.	424.	425.	424.	426.	427.
948 7	450.00	439.	441.	443.	443.	446.	447.	445.	447.	443.	447.	449.	451.
953 7	503.67	487.	491.	495.	497.	500.	501.	503.	504.	504.	504.	504.	504.
957 3	531.00	523.	524.	525.	518.	525.	524.	527.	528.	529.	529.	531.	531.
960 4	544.67	535.	536.	537.	532.	533.	532.	535.	541.	542.	544.	544.	546.

Output Table 4A-5S

ATE OF AK	STANDING AT PEAK	-11 MO	-10 MO	ABSOLUTE CHANGE -9 MO	-8 MO	-7 MO	-6 MO	ONE YEAR BEFORE PEAK -5 MO	-4 MO	-3 MO	-2 MO	-1 MO	0 MO
29 8	332.00	0.0	0.0	0.0	0.0	-6.0	-6.0	-4.0	-3.0	-2.0	-1.0	0.0	1.0
37 7	320.67	-17.7	-15.7	-14.7	-11.7	-8.7	-7.7	-5.7	-3.7	-2.7	-0.7	-0.7	0.3
43 11	426.33	-8.3	-5.3	-3.3	-2.3	-2.3	-2.3	-0.3	-2.3	-1.3	-2.3	-0.3	0.7
48 7	450.00	-11.0	-9.0	-7.0	-7.0	-4.0	-3.0	-5.0	-3.0	-7.0	-3.0	-1.0	1.0
53 7	503.67	-16.7	-12.7	-8.7	-6.7	-3.7	-2.7	-0.7	0.3	0.3	0.3	0.3	0.3
57 3	531.00	-8.0	-7.0	-6.0	-13.0	-6.0	-7.0	-4.0	-3.0	-2.0	-2.0	0.0	0.0
60 4	544.67	-9.7	-8.7	-7.7	-12.7	-11.7	-12.7	-9.7	-3.7	-2.7	-0.7	-0.7	1.3
OTAL		-71.3	-58.3	-47.3	-53.3	-42.3	-41.3	-29.3	-18.3	-17.3	-9.3	-2.3	4.7
VERAGE		-11.9	-9.7	-7.9	-8.9	-6.0	-5.9	-4.2	-2.6	-2.5	-1.3	-0.3	0.7
VE DEVIATION		3.5	3.0	2.5	3.6	2.4	2.8	2.2	0.9	1.4	1.0	0.4	0.4

OMIT THE FOLLOWING CYCLES

TAL		-71.3	-58.3	-47.3	-53.3	-42.3	-41.3	-29.3	-18.3	-17.3	-9.3	-2.3	4.7
ERAGE		-11.9	-9.7	-7.9	-8.9	-6.0	-5.9	-4.2	-2.6	-2.5	-1.3	-0.3	0.7
E DEVIATION		3.5	3.0	2.5	3.6	2.4	2.8	2.2	0.9	1.4	1.0	0.4	0.4

Output Table 4A-6S

TE OF K	STANDING AT PEAK	-11 MO	-10 MO	RELATIVE STANDING -9 MO	-8 MO	-7 MO	-6 MO	ONE YEAR BEFORE PEAK -5 MO	-4 MO	-3 MO	-2 MO	-1 MO	0 MO
9 8	332.00	0.0	0.0	0.0	0.0	98.2	98.2	98.8	99.1	99.4	99.7	100.0	100.3
7 7	320.67	94.5	95.1	95.4	96.4	97.3	97.6	98.2	98.9	99.2	99.8	99.8	100.1
3 11	426.33	98.0	98.7	99.2	99.5	99.5	99.5	99.9	99.5	99.7	99.5	99.9	100.2
3 7	450.00	97.6	98.0	98.4	98.4	99.1	99.3	98.9	99.3	98.4	99.3	99.8	100.2
3 7	503.67	96.7	97.5	98.3	98.7	99.3	99.5	99.9	100.1	100.1	100.1	100.1	100.1
7 3	531.00	98.5	98.7	98.9	97.6	98.9	98.7	99.2	99.4	99.6	99.6	100.0	100.0
0 4	544.67	98.2	98.4	98.6	97.7	97.9	97.7	98.2	99.3	99.5	99.9	99.9	100.2
TAL		583.5	586.4	588.8	588.2	690.1	690.4	693.2	695.6	695.9	697.8	699.4	701.1
ERAGE		97.3	97.7	98.1	98.0	98.6	98.6	99.0	99.4	99.4	99.7	99.9	100.2
E DEVIATION		1.1	1.0	0.9	0.8	0.7	0.7	0.6	0.2	0.4	0.2	0.1	0.1

OMIT THE FOLLOWING CYCLES

AL		583.5	586.4	588.8	588.2	690.1	690.4	693.2	695.6	695.9	697.8	699.4	701.1
RAGE		97.3	97.7	98.1	98.0	98.6	98.6	99.0	99.4	99.4	99.7	99.9	100.2
DEVIATION		1.1	1.0	0.9	0.8	0.7	0.7	0.6	0.2	0.4	0.2	0.1	0.1

Output Table 4A-4S

EMPLOYEES IN NONAG ESTABLISHMENTS

HUNDRED THOUSAND PERSONS 8268

SPECIFIC ANALYSIS

DATE OF PEAK	STANDING AT PEAK	+1 MO	+2 MO	STANDINGS STANDING +3 MO	+4 MO	+5 MO	+6 MO	FIRST YEAR AFTER PEAK +7 MO	+8 MO	+9 MO	+10 MO	+11 MO	+12 MO
1929 8	332.00	331.	330.	327.	323.	319.	317.	314.	313.	311.	308.	304.	300
1937 7	320.67	321.	320.	317.	312.	305.	299.	296.	294.	292.	288.	287.	287
1943 11	426.33	426.	425.	424.	422.	419.	418.	417.	417.	417.	416.	416.	416
1948 7	450.00	450.	451.	451.	451.	451.	446.	444.	442.	441.	438.	436.	435
1953 7	503.67	503.	502.	501.	498.	497.	494.	493.	491.	490.	489.	488.	487
1957 3	531.00	531.	530.	530.	530.	530.	528.	527.	525.	523.	521.	515.	51[illegible]
1960 4	544.67	544.	543.	542.	542.	541.	540.	539.	536.	535.	534.	535.	53[illegible]

Output Table 4A-5S

DATE OF PEAK	STANDING AT PEAK	+1 MO	+2 MO	ABSOLUTE CHANGE +3 MO	+4 MO	+5 MO	+6 MO	FIRST YEAR AFTER PEAK +7 MO	+8 MO	+9 MO	+10 MO	+11 MO	+12 MO
1929 8	332.00	-1.0	-2.0	-5.0	-9.0	-13.0	-15.0	-18.0	-19.0	-21.0	-24.0	-28.0	-32.
1937 7	320.67	0.3	-0.7	-3.7	-8.7	-15.7	-21.7	-24.7	-26.7	-28.7	-32.7	-33.7*	-33.
1943 11	426.33	-0.3	-1.3	-2.3	-4.3	-7.3	-8.3	-9.3	-9.3	-9.3	-10.3	-10.3	-10.
1948 7	450.00	0.0	1.0	1.0	1.0	1.0	-4.0	-6.0	-8.0	-9.0	-12.0	-14.0	-15.
1953 7	503.67	-0.7	-1.7	-2.7	-5.7	-6.7	-9.7	-10.7	-12.7	-13.7	-14.7	-15.7	-16.
1957 3	531.00	0.0	-1.0	-1.0	-1.0	-1.0	-3.0	-4.0	-6.0	-8.0	-10.0	-16.0	-19.
1960 4	544.67	-0.7	-1.7	-2.7	-2.7	-3.7	-4.7	-5.7	-8.7	-9.7	-10.7*	-9.7	-9.
TOTAL		-2.3	-7.3	-16.3	-30.3	-46.3	-66.3	-78.3	-90.3	-99.3	-114.3	-127.3	-136.
AVERAGE		-0.3	-1.0	-2.3	-4.3	-6.6	-9.5	-11.2	-12.9	-14.2	-16.3	-18.2	-19.
AVE DEVIATION		0.4	0.7	1.3	3.0	4.6	5.1	5.8	5.7	6.1	6.9	7.2	7.

OMIT THE FOLLOWING CYCLES

		+1 MO	+2 MO	+3 MO	+4 MO	+5 MO	+6 MO	+7 MO	+8 MO	+9 MO	+10 MO	+11 MO	+12 MO
TOTAL		-2.3	-7.3	-16.3	-30.3	-46.3	-66.3	-78.3	-90.3	-99.3	-114.3	-127.3	-136
AVERAGE		-0.3	-1.0	-2.3	-4.3	-6.6	-9.5	-11.2	-12.9	-14.2	-16.3	-18.2	-19
AVE DEVIATION		0.4	0.7	1.3	3.0	4.6	5.1	5.8	5.7	6.1	6.9	7.2	7

Output Table 4A-6S

DATE OF PEAK	STANDING AT PEAK	+1 MO	+2 MO	RELATIVE STANDING +3 MO	+4 MO	+5 MO	+6 MO	FIRST YEAR AFTER PEAK +7 MO	+8 MO	+9 MO	+10 MO	+11 MO	+12
1929 8	332.00	99.7	99.4	98.5	97.3	96.1	95.5	94.6	94.3	93.7	92.8	91.6	90
1937 7	320.67	100.1	99.8	98.9	97.3	95.1	93.2	92.3	91.7	91.1	89.8	89.5*	89
1943 11	426.33	99.9	99.7	99.5	99.0	98.3	98.0	97.8	97.8	97.8	97.6	97.6	97
1948 7	450.00	100.0	100.2	100.2	100.2	100.2	99.1	98.7	98.2	98.0	97.3	96.9	96
1953 7	503.67	99.9	99.7	99.5	98.9	98.7	98.1	97.9	97.5	97.3	97.1	96.9	96
1957 3	531.00	100.0	99.8	99.8	99.8	99.8	99.4	99.2	98.9	98.5	98.1	97.0	96
1960 4	544.67	99.9	99.7	99.5	99.5	99.3	99.1	99.0	98.4	98.2	98.0*	98.2	98
TOTAL		699.5	698.3	695.8	692.0	687.5	682.5	679.5	676.8	674.6	670.7	667.6	665
AVERAGE		99.9	99.8	99.4	98.9	98.2	97.5	97.1	96.7	96.4	95.8	95.4	95
AVE DEVIATION		0.1	0.2	0.4	0.9	1.5	1.8	2.1	2.1	2.3	2.6	2.8	2

OMIT THE FOLLOWING CYCLES

		+1 MO	+2 MO	+3 MO	+4 MO	+5 MO	+6 MO	+7 MO	+8 MO	+9 MO	+10 MO	+11 MO	+12
TOTAL		699.5	698.3	695.8	692.0	687.5	682.5	679.5	676.8	674.6	670.7	667.6	665
AVERAGE		99.9	99.8	99.4	98.9	98.2	97.5	97.1	96.7	96.4	95.8	95.4	95
AVE DEVIATION		0.1	0.2	0.4	0.9	1.5	1.8	2.1	2.1	2.3	2.6	2.8	2

APPENDIX TO CHAPTER 4

B

SAMPLE RUN, RECESSION ANALYSIS

Output Table 4B-1S

NBER RECESSION-RECOVERY ANALYSIS

SPECIFIC ANALYSIS

EMPLOYEES IN NONAG ESTABLISHMENTS

HUNDRED THOUSAND PERSONS

8268

YEAR	JAN	FEB	MAR	APR	MAY	JUNE	JULY	AUG	SEPT	OCT	NOV	DEC
1929	326.0	326.0	328.0	329.0	330.0	331.0	332.0	333.0	331.0	330.0	327.0	323.
1930	319.0	317.0	314.0	313.0	311.0	308.0	304.0	300.0	297.0	294.0	291.0	289.
1931	286.0	285.0	283.0	282.0	280.0	277.0	275.0	272.0	268.0	265.0	262.0	260.
1961	535.0	534.0	535.0	535.0	537.0	539.0	541.0	542.0	542.0	543.0	546.0	547.
1962	547.0	550.0	552.0	554.0	555.0	556.0	557.0	557.0	558.0	558.0	559.0	559.
1963	559.0	560.0	562.0	564.0	565.0	566.0	568.0	568.0	569.0	571.0	571.0	573.

Output Table 4B-2S

ABSOLUTE CHANGES BETWEEN CYCLICAL TURNS

PEAK	TROUGH	PEAK	PEAK	TROUGH	PEAK	FALL	RISE
1929 8	1933 3	1937 7	332.0	230.0	320.7	-102.0	90.7
1937 7	1938 6	1943 11	320.7	287.3	426.3	-33.3	139.0
1943 11	1945 9	1948 7	426.3	390.7	450.0	-35.7	59.3
1948 7	1949 10	1953 7	450.0	432.3	503.7	-17.7	71.3
1953 7	1954 8	1957 3	503.7	487.0	531.0	-16.7	44.0
1957 3	1958 5	1960 4	531.0	508.7	544.7	-22.3	36.0
1960 4	1961 2		544.7	534.7		-10.0	
TOTAL						-237.7	440.3
AVERAGE						-34.0	73.4
AVERAGE DEVIATIONS						19.9	27.6

Output Table 4B-3S

RELATIVE CHANGES BETWEEN CYCLICAL TURNS

PEAK	TROUGH	PEAK	PEAK	TROUGH	PEAK	FALL	RISE
1929 8	1933 3	1937 7	332.0	230.0	320.7	-30.7	39.4
1937 7	1938 6	1943 11	320.7	287.3	426.3	-10.4	48.4
1943 11	1945 9	1948 7	426.3	390.7	450.0	-8.4	15.2
1948 7	1949 10	1953 7	450.0	432.3	503.7	-3.9	16.5
1953 7	1954 8	1957 3	503.7	487.0	531.0	-3.3	9.0
1957 3	1958 5	1960 4	531.0	508.7	544.7	-4.2	7.1
1960 4	1961 2		544.7	534.7		-1.8	
TOTAL						-62.8	135.6
AVERAGE						-9.0	22.6
AVERAGE DEVIATIONS						6.6	14.2

Output Table 4B-4R

EMPLOYEES IN NONAG ESTABLISHMENTS

HUNDRED THOUSAND PERSONS 8268

REFERENCE ANALYSIS

DATE OF ROUGH	STANDING AT TROUGH	-11 MO	-10 MO	STANDINGS STANDING -9 MO	-8 MO	-7 MO	-6 MO	ONE YEAR BEFORE TROUGH -5 MO	-4 MO	-3 MO	-2 MO	-1 MO	0. MO
933 3	230.00	246.	242.	238.	234.	233.	235.	237.	237.	235.	234.	232.	228.
938 6	287.33	321.	321.	320.	317.	312.	305.	299.	296.	294.	292.	288.	287.
945 10	385.67	416.	417.	418.	418.	417.	413.	411.	409.	406.	403.	384.	385.
949 10	432.33	451.	451.	446.	444.	442.	441.	438.	436.	435.	435.	437.	428.
954 8	487.00	502.	501.	498.	497.	494.	493.	491.	490.	489.	488.	487.	487.
58 4	509.67	530.	530.	530.	530.	528.	527.	525.	523.	521.	515.	512.	509.

Output Table 4B-5R

ATE OF OUGH	STANDING AT TROUGH	-11 MO	-10 MO	ABSOLUTE CHANGE -9 MO	-8 MO	-7 MO	-6 MO	ONE YEAR BEFORE TROUGH -5 MO	-4 MO	-3 MO	-2 MO	-1 MO	0 MO
33 3	230.00	16.0	12.0	8.0	4.0	3.0	5.0	7.0	7.0	5.0	4.0	2.0	-2.0
38 6	287.33	33.7	33.7	32.7	29.7	24.7	17.7	11.7	8.7	6.7	4.7	0.7	-0.3
45 10	385.67	30.3	31.3	32.3	32.3	31.3	27.3	25.3	23.3	20.3	17.3	-1.7	-0.7
49 10	432.33	18.7	18.7	13.7	11.7	9.7	8.7	5.7	3.7	2.7	2.7	4.7	-4.3
54 8	487.00	15.0	14.0	11.0	10.0	7.0	6.0	4.0	3.0	2.0	1.0	0.0	0.0
58 4	509.67	20.3	20.3	20.3	20.3	18.3	17.3	15.3	13.3	11.3	5.3	2.3	-0.7
TAL		134.0	130.0	118.0	108.0	94.0	82.0	69.0	59.0	48.0	35.0	8.0	-8.0
VERAGE		22.3	21.7	19.7	18.0	15.7	13.7	11.5	9.8	8.0	5.8	1.3	-1.3
E DEVIATION		6.4	7.2	8.8	9.4	9.1	7.1	5.9	5.7	5.2	3.8	1.7	1.2

OMIT THE FOLLOWING CYCLES

1945 10 1948 11 1949 10

		-11 MO	-10 MO	-9 MO	-8 MO	-7 MO	-6 MO	-5 MO	-4 MO	-3 MO	-2 MO	-1 MO	0 MO
TAL		103.7	98.7	85.7	75.7	62.7	54.7	43.7	35.7	27.7	17.7	9.7	-7.3
ERAGE		20.7	19.7	17.1	15.1	12.5	10.9	8.7	7.1	5.5	3.5	1.9	-1.5
E DEVIATION		5.2	5.8	7.5	7.9	7.2	5.3	3.8	3.1	2.8	1.4	1.3	1.4

Output Table 4B-6R

TE OF UGH	STANDING AT TROUGH	-11 MO	-10 MO	RELATIVE STANDING -9 MO	-8 MO	-7 MO	-6 MO	ONE YEAR BEFORE TROUGH -5 MO	-4 MO	-3 MO	-2 MO	-1 MO	0 MO
3 3	230.00	107.0	105.2	103.5	101.7	101.3	102.2	103.0	103.0	102.2	101.7	100.9	99.1
8 6	287.33	111.7	111.7	111.4	110.3	108.6	106.1	104.1	103.0	102.3	101.6	100.2	99.9
5 10	385.67	107.9	108.1	108.4	108.4	108.1	107.1	106.6	106.1	105.3	104.5	99.6	99.8
9 10	432.33	104.3	104.3	103.2	102.7	102.2	102.0	101.3	100.8	100.6	100.6	101.1	99.0
4 8	487.00	103.1	102.9	102.3	102.1	101.4	101.2	100.8	100.6	100.4	100.2	100.0	100.0
8 4	509.67	104.0	104.0	104.0	104.0	103.6	103.4	103.0	102.6	102.2	101.0	100.5	99.9
AL		637.9	636.2	632.6	629.2	625.3	622.0	618.8	616.2	613.0	609.7	602.2	597.7
RAGE		106.3	106.0	105.4	104.9	104.2	103.7	103.1	102.7	102.2	101.6	100.4	99.6
DEVIATION		2.5	2.6	3.0	3.0	2.8	2.0	1.5	1.3	1.1	1.0	0.4	0.4

OMIT THE FOLLOWING CYCLES

1945 10 1948 11 1949 10

		-11 MO	-10 MO	-9 MO	-8 MO	-7 MO	-6 MO	-5 MO	-4 MO	-3 MO	-2 MO	-1 MO	0 MO
AL		530.1	528.1	524.3	520.8	517.2	515.0	512.2	510.1	507.7	505.2	502.6	497.9
RAGE		106.0	105.6	104.9	104.2	103.4	103.0	102.4	102.0	101.5	101.0	100.5	99.6
DEVIATION		2.7	2.4	2.6	2.5	2.1	1.4	1.1	1.0	0.8	0.5	0.4	0.4

Output Table 4B-4R

EMPLOYEES IN NONAG ESTABLISHMENTS

HUNDRED THOUSAND PERSONS 8268

REFERENCE ANALYSIS

STANDINGS

DATE OF TROUGH	STANDING AT TROUGH	+1 MO	+2 MO	STANDING +3 MO	+4 MO	+5 MO	+6 MO	FIRST YEAR AFTER TROUGH +7 MO	+8 MO	+9 MO	+10 MO	+11 MO	+12 M
1933 3	230.00	230.	233.	239.	245.	252.	256.	259.	259.	258.	259.	263.	26
1938 6	287.33	287.	289.	292.	293.	297.	299.	298.	299.	302.	300.	302.	30
1945 10	385.67	388.	390.	397.	392.	402.	408.	413.	416.	420.	425.	428.	43
1949 10	432.33	432.	435.	435.	432.	439.	443.	446.	450.	454.	461.	463.	46
1954 8	487.00	487.	488.	491.	493.	494.	496.	499.	501.	505.	507.	509.	50
1958 4	509.67	508.	509.	509.	512.	514.	514.	519.	520.	524.	526.	529.	53

Output Table 4B-5R

DATE OF TROUGH	STANDING AT TROUGH	+1 MO	+2 MO	ABSOLUTE CHANGE +3 MO	+4 MO	+5 MO	+6 MO	FIRST YEAR AFTER TROUGH +7 MO	+8 MO	+9 MO	+10 MO	+11 MO	+12
1933 3	230.00	0.0	3.0	9.0	15.0	22.0	26.0	29.0	29.0	28.0	29.0	33.0	37
1938 6	287.33	-0.3	1.7	4.7	5.7	9.7	11.7	10.7	11.7	14.7	12.7	14.7	16
1945 10	385.67	2.3	4.3	11.3	6.3	16.3	22.3	27.3	30.3	34.3	39.3	42.3	44
1949 10	432.33	-0.3	2.7	2.7	-0.3	6.7	10.7	13.7	17.7	21.7	28.7	30.7	32
1954 8	487.00	0.0	1.0	4.0	6.0	7.0	9.0	12.0	14.0	18.0	20.0	22.0	22
1958 4	509.67	-1.7	-0.7	-0.7	2.3	4.3	4.3	9.3	10.3	14.3	16.3	19.3	22
TOTAL		0.0	12.0	31.0	35.0	66.0	84.0	102.0	113.0	131.0	146.0	162.0	175
AVERAGE		0.0	2.0	5.2	5.8	11.0	14.0	17.0	18.8	21.8	24.3	27.0	29
AVE DEVIATION		0.8	1.3	3.3	3.3	5.4	6.8	7.4	7.2	6.2	8.0	8.3	[illegible]

OMIT THE FOLLOWING CYCLES

1945 10 1948 11 1949 10

		+1 MO	+2 MO	+3 MO	+4 MO	+5 MO	+6 MO	+7 MO	+8 MO	+9 MO	+10 MO	+11 MO	+12
TOTAL		-2.3	7.7	19.7	28.7	49.7	61.7	74.7	82.7	96.7	106.7	119.7	13
AVERAGE		-0.5	1.5	3.9	5.7	9.9	12.3	14.9	16.5	19.3	21.3	23.9	[illegible]
AVE DEVIATION		0.5	1.1	2.3	3.8	4.8	5.5	5.6	5.4	4.4	6.0	6.3	

Output Table 4B-6R

DATE OF TROUGH	STANDING AT TROUGH	+1 MO	+2 MO	RELATIVE STANDING +3 MO	+4 MO	+5 MO	+6 MO	FIRST YEAR AFTER TROUGH +7 MO	+8 MO	+9 MO	+10 MO	+11 MO	+12
1933 3	230.00	100.0	101.3	103.9	106.5	109.6	111.3	112.6	112.6	112.2	112.6	114.3	1[illegible]
1938 6	287.33	99.9	100.6	101.6	102.0	103.4	104.1	103.7	104.1	105.1	104.4	105.1	10
1945 10	385.67	100.6	101.1	102.9	101.6	104.2	105.8	107.1	107.9	108.9	110.2	111.0	1[illegible]
1949 10	432.33	99.9	100.6	100.6	99.9	101.5	102.5	103.2	104.1	105.0	106.6	107.1	10
1954 8	487.00	100.0	100.2	100.8	101.2	101.4	101.8	102.5	102.9	103.7	104.1	104.5	10
1958 4	509.67	99.7	99.9	99.9	100.5	100.9	100.9	101.8	102.0	102.8	103.2	103.8	10
TOTAL		600.1	603.7	609.8	611.7	621.0	626.3	630.9	633.5	637.7	641.2	645.8	6[illegible]
AVERAGE		100.0	100.6	101.6	102.0	103.5	104.4	105.1	105.6	106.3	106.9	107.6	1[illegible]
AVE DEVIATION		0.2	0.4	1.2	1.5	2.3	2.8	3.1	3.1	2.8	3.0	3.3	

OMIT THE FOLLOWING CYCLES

1945 10 1948 11 1949 10

		+1 MO	+2 MO	+3 MO	+4 MO	+5 MO	+6 MO	+7 MO	+8 MO	+9 MO	+10 MO	+11 MO	+12
TOTAL		499.5	502.6	506.8	510.1	516.8	520.5	523.8	525.7	528.8	531.0	534.9	[illegible]
AVERAGE		99.9	100.5	101.4	102.0	103.4	104.1	104.8	105.1	105.8	106.2	107.0	[illegible]
AVE DEVIATION		0.1	0.4	1.1	1.8	2.5	2.9	3.1	3.0	2.6	2.7	3.0	

Output Table 4B-4R

)YEES IN NONAG ESTABLISHMENTS

HUNDRED THOUSAND PERSONS 8268

RENCE ANALYSIS

STANDINGS

OF H	STANDING AT PEAK	-11 MO	-10 MO	STANDING -9 MO	-8 MO	-7 MO	-6 MO	ONE YEAR BEFORE TROUGH -5 MO	-4 MO	-3 MO	-2 MO	-1 MO	0 MO
3	332.00	246.	242.	238.	234.	233.	235.	237.	237.	235.	234.	232.	228.
6	319.33	321.	321.	320.	317.	312.	305.	299.	296.	294.	292.	288.	287.
10	417.67	416.	417.	418.	418.	417.	413.	411.	409.	406.	403.	384.	385.
10	451.00	451.	451.	446.	444.	442.	441.	438.	436.	435.	435.	437.	428.
8	503.67	502.	501.	498.	497.	494.	493.	491.	490.	489.	488.	487.	487.
4	530.00	530.	530.	530.	530.	528.	527.	525.	523.	521.	515.	512.	509.

Output Table 4B-7R

OF H	STANDING AT PEAK	-11 MO	-10 MO	ABSOLUTE CHANGE -9 MO	-8 MO	-7 MO	-6 MO	ONE YEAR BEFORE TROUGH -5 MO	-4 MO	-3 MO	-2 MO	-1 MO	0 MO
3	332.00	-86.0	-90.0	-94.0	-98.0	-99.0	-97.0	-95.0	-95.0	-97.0	-98.0	-100.0	-104.0
6	319.33	1.7	1.7	0.7	-2.3	-7.3	-14.3	-20.3	-23.3	-25.3	-27.3	-31.3	-32.3
10	417.67	-1.7	-0.7	0.3	0.3	-0.7	-4.7	-6.7	-8.7	-11.7	-14.7	-33.7	-32.7
10	451.00	0.0	0.0	-5.0	-7.0	-9.0	-10.0	-13.0	-15.0	-16.0	-16.0	-14.0	-23.0
8	503.67	-1.7	-2.7	-5.7	-6.7	-9.7	-10.7	-12.7	-13.7	-14.7	-15.7	-16.7	-16.7
4	530.00	0.0	0.0	0.0	0.0	-2.0	-3.0	-5.0	-7.0	-9.0	-15.0	-18.0	-21.0
L		-87.7	-91.7	-103.7	-113.7	-127.7	-139.7	-152.7	-162.7	-173.7	-186.7	-213.7	-229.7
AGE		-14.6	-15.3	-17.3	-18.9	-21.3	-23.3	-25.4	-27.1	-28.9	-31.1	-35.6	-38.3
DEVIATION		23.8	24.9	25.6	26.4	25.9	24.6	23.2	22.6	22.7	22.3	21.5	21.9

OMIT THE FOLLOWING CYCLES

1945 10 1948 11 1949 10

	-11 MO	-10 MO	-9 MO	-8 MO	-7 MO	-6 MO	-5 MO	-4 MO	-3 MO	-2 MO	-1 MO	0 MO
L	-86.0	-91.0	-104.0	-114.0	-127.0	-135.0	-146.0	-154.0	-162.0	-172.0	-180.0	-197.0
AGE	-17.2	-18.2	-20.8	-22.8	-25.4	-27.0	-29.2	-30.8	-32.4	-34.4	-36.0	-39.4
DEVIATION	27.5	28.7	29.3	30.1	29.4	28.0	26.3	25.7	25.8	25.4	25.6	25.8

Output Table 4B-8R

OF H	STANDING AT PEAK	-11 MO	-10 MO	RELATIVE STANDING -9 MO	-8 MO	-7 MO	-6 MO	ONE YEAR BEFORE TROUGH -5 MO	-4 MO	-3 MO	-2 MO	-1 MO	0 MO
3	332.00	74.1	72.9	71.7	70.5	70.2	70.8	71.4	71.4	70.8	70.5	69.9	68.7
6	319.33	100.5	100.5	100.2	99.3	97.7	95.5	93.6	92.7	92.1	91.4	90.2	89.9
10	417.67	99.6	99.8	100.1	100.1	99.8	98.9	98.4	97.9	97.2	96.5	91.9	92.2
10	451.00	100.0	100.0	98.9	98.4	98.0	97.8	97.1	96.7	96.5	96.5	96.9	94.9
8	503.67	99.7	99.5	98.9	98.7	98.1	97.9	97.5	97.3	97.1	96.9	96.7	96.7
4	530.00	100.0	100.0	100.0	100.0	99.6	99.4	99.1	98.7	98.3	97.2	96.6	96.0
L		573.9	572.7	569.7	567.0	563.4	560.3	557.1	554.6	551.9	548.9	542.2	538.4
AGE		95.6	95.5	95.0	94.5	93.9	93.4	92.8	92.4	92.0	91.5	90.4	89.7
DEVIATION		7.2	7.5	7.8	8.0	7.9	7.5	7.2	7.0	7.1	7.0	6.9	7.0

OMIT THE FOLLOWING CYCLES

1945 10 1948 11 1949 10

	-11 MO	-10 MO	-9 MO	-8 MO	-7 MO	-6 MO	-5 MO	-4 MO	-3 MO	-2 MO	-1 MO	0 MO
L	474.3	472.9	469.7	466.9	463.6	461.4	458.7	456.7	454.7	452.4	450.3	446.2
AGE	94.9	94.6	93.9	93.4	92.7	92.3	91.7	91.3	90.9	90.5	90.1	89.2
DEVIATION	8.3	8.7	8.9	9.2	9.0	8.6	8.1	8.0	8.1	8.0	8.1	8.2

Output Table 4B-4R

EMPLOYEES IN NONAG ESTABLISHMENTS

HUNDRED THOUSAND PERSONS

8268

REFERENCE ANALYSIS

STANDINGS

DATE OF TROUGH	STANDING AT PEAK	+1 MO	+2 MO	STANDING +3 MO	+4 MO	+5 MO	+6 MO	FIRST YEAR AFTER TROUGH +7 MO	+8 MO	+9 MO	+10 MO	+11 MO
1933 3	332.00	230.	233.	239.	245.	252.	256.	259.	259.	258.	259.	263.
1938 6	319.33	287.	289.	292.	293.	297.	299.	298.	299.	302.	300.	302.
1945 10	417.67	388.	390.	397.	392.	402.	408.	413.	416.	420.	425.	428.
1949 10	451.00	432.	435.	435.	432.	439.	443.	446.	450.	454.	461.	463.
1954 8	503.67	487.	488.	491.	493.	494.	496.	499.	501.	505.	507.	509.
1958 4	530.00	508.	509.	509.	512.	514.	514.	519.	520.	524.	526.	529.

Output Table 4B-7R

DATE OF TROUGH	STANDING AT PEAK	+1 MO	+2 MO	ABSOLUTE CHANGE +3 MO	+4 MO	+5 MO	+6 MO	FIRST YEAR AFTER TROUGH +7 MO	+8 MO	+9 MO	+10 MO	+11 MO
1933 3	332.00	-102.0	-99.0	-93.0	-87.0	-80.0	-76.0	-73.0	-73.0	-74.0	-73.0	-69.0
1938 6	319.33	-32.3	-30.3	-27.3	-26.3	-22.3	-20.3	-21.3	-20.3	-17.3	-19.3	-17.3
1945 10	417.67	-29.7	-27.7	-20.7	-25.7	-15.7	-9.7	-4.7	-1.7	2.3	7.3	10.3
1949 10	451.00	-19.0	-16.0	-16.0	-19.0	-12.0	-8.0	-5.0	-1.0	3.0	10.0	12.0
1954 8	503.67	-16.7	-15.7	-12.7	-10.7	-9.7	-7.7	-4.7	-2.7	1.3	3.3	5.3
1958 4	530.00	-22.0	-21.0	-21.0	-18.0	-16.0	-16.0	-11.0	-10.0	-6.0	-4.0	-1.0
TOTAL		-221.7	-209.7	-190.7	-186.7	-155.7	-137.7	-119.7	-108.7	-90.7	-75.7	-59.7
AVERAGE		-36.9	-34.9	-31.8	-31.1	-25.9	-22.9	-19.9	-18.1	-15.1	-12.6	-9.9
AVE DEVIATION		21.7	21.4	20.4	18.6	18.0	17.7	18.1	19.0	20.4	22.4	22.1

OMIT THE FOLLOWING CYCLES

1945 10 1948 11 1949 10

	+1 MO	+2 MO	+3 MO	+4 MO	+5 MO	+6 MO	+7 MO	+8 MO	+9 MO	+10 MO	+11 MO
TOTAL	-192.0	-182.0	-170.0	-161.0	-140.0	-128.0	-115.0	-107.0	-93.0	-83.0	-70.0
AVERAGE	-38.4	-36.4	-34.0	-32.2	-28.0	-25.6	-23.0	-21.4	-18.6	-16.6	-14.0
AVE DEVIATION	25.4	25.0	23.6	21.9	20.8	20.2	20.0	20.6	22.2	23.7	23.3

Output Table 4B-8R

DATE OF TROUGH	STANDING AT PEAK	+1 MO	+2 MO	RELATIVE STANDING +3 MO	+4 MO	+5 MO	+6 MO	FIRST YEAR AFTER TROUGH +7 MO	+8 MO	+9 MO	+10 MO	+11 MO
1933 3	332.00	69.3	70.2	72.0	73.8	75.9	77.1	78.0	78.0	77.7	78.0	79.2
1938 6	319.33	89.9	90.5	91.4	91.8	93.0	93.6	93.3	93.6	94.6	93.9	94.6
1945 10	417.67	92.9	93.4	95.1	93.9	96.2	97.7	98.9	99.6	100.6	101.8	102.5
1949 10	451.00	95.8	96.5	96.5	95.8	97.3	98.2	98.9	99.8	100.7	102.2	102.7
1954 8	503.67	96.7	96.9	97.5	97.9	98.1	98.5	99.1	99.5	100.3	100.7	101.1
1958 4	530.00	95.8	96.0	96.0	96.6	97.0	97.0	97.9	98.1	98.9	99.2	99.8
TOTAL		540.4	543.4	548.5	549.7	557.6	562.1	566.1	568.6	572.6	575.8	579.8
AVERAGE		90.1	90.6	91.4	91.6	92.9	93.7	94.4	94.8	95.4	96.0	96.6
AVE DEVIATION		7.0	6.8	6.5	5.9	5.7	5.5	5.8	6.0	6.2	6.7	6.5

OMIT THE FOLLOWING CYCLES

1945 10 1948 11 1949 10

	+1 MO	+2 MO	+3 MO	+4 MO	+5 MO	+6 MO	+7 MO	+8 MO	+9 MO	+10 MO	+11 MO
TOTAL	447.5	450.1	453.4	455.8	461.3	464.4	467.2	469.0	472.1	474.1	477.3
AVERAGE	89.5	90.0	90.7	91.2	92.3	92.9	93.4	93.8	94.4	94.8	95.5
AVE DEVIATION	8.1	7.9	7.5	6.9	6.5	6.3	6.2	6.4	6.7	7.1	6.9

Output Table 4B-4S

MPLOYEES IN NONAG ESTABLISHMENTS

HUNDRED THOUSAND PERSONS 8268

PECIFIC ANALYSIS

ATE OF OUGH	STANDING AT TROUGH	-11 MO	-10 MO	STANDINGS STANDING -9 MO	-8 MO	-7 MO	-6 MO	ONE YEAR BEFORE TROUGH -5 MO	-4 MO	-3 MO	-2 MO	-1 MO	0 MO
33 3	230.00	246.	242.	238.	234.	233.	235.	237.	237.	235.	234.	232.	228.
38 6	287.33	321.	321.	320.	317.	312.	305.	299.	296.	294.	292.	288.	287.
45 9	390.67	416.	416.	417.	418.	418.	417.	413.	411.	409.	406.	403.	384.
9 10	432.33	451.	451.	446.	444.	442.	441.	438.	436.	435.	435.	437.	428.
4 8	487.00	502.	501.	498.	497.	494.	493.	491.	490.	489.	488.	487.	487.
8 5	508.67	530.	530.	530.	528.	527.	525.	523.	521.	515.	512.	509.	508.
1 2	534.67	544.	546.	544.	543.	542.	542.	541.	540.	539.	536.	535.	534.

Output Table 4B-5S

TE OF UGH	STANDING AT TROUGH	-11 MO	-10 MO	ABSOLUTE CHANGE -9 MO	-8 MO	-7 MO	-6 MO	ONE YEAR BEFORE TROUGH -5 MO	-4 MO	-3 MO	-2 MO	-1 MO	0 MO
3 3	230.00	16.0	12.0	8.0	4.0	3.0	5.0	7.0	7.0	5.0	4.0	2.0	-2.0
8 6	287.33	33.7	33.7	32.7	29.7	24.7	17.7	11.7	8.7	6.7	4.7	0.7	-0.3
5 9	390.67	25.3	25.3	26.3	27.3	27.3	26.3	22.3	20.3	18.3	15.3	12.3	-6.7
9 10	432.33	18.7	18.7	13.7	11.7	9.7	8.7	5.7	3.7	2.7	2.7	4.7	-4.3
4 8	487.00	15.0	14.0	11.0	10.0	7.0	6.0	4.0	3.0	2.0	1.0	0.0	0.0
8 5	508.67	21.3	21.3	21.3	19.3	18.3	16.3	14.3	12.3	6.3	3.3	0.3	-0.7
1 2	534.67	9.3	11.3	9.3	8.3	7.3	7.3	6.3	5.3	4.3	1.3	0.3	-0.7
TAL		139.3	136.3	122.3	110.3	97.3	87.3	71.3	60.3	45.3	32.3	20.3	-14.7
ERAGE		19.9	19.5	17.5	15.8	13.9	12.5	10.2	8.6	6.5	4.6	2.9	-2.1
E DEVIATION		5.9	6.3	8.0	8.3	8.2	6.5	5.1	4.4	3.4	3.1	3.2	1.9

OMIT THE FOLLOWING CYCLES

1945 9 1948 7 1949 10

		-11 MO	-10 MO	-9 MO	-8 MO	-7 MO	-6 MO	-5 MO	-4 MO	-3 MO	-2 MO	-1 MO	0 MO
AL		114.0	111.0	96.0	83.0	70.0	61.0	49.0	40.0	27.0	17.0	8.0	-8.0
RAGE		19.0	18.5	16.0	13.8	11.7	10.2	8.2	6.7	4.5	2.8	1.3	-1.3
DEVIATION		5.7	6.1	7.3	7.1	6.6	4.6	3.2	2.7	1.5	1.2	1.3	1.2

Output Table 4B-6S

E OF GH	STANDING AT TROUGH	-11 MO	-10 MO	RELATIVE STANDING -9 MO	-8 MO	-7 MO	-6 MO	ONE YEAR BEFORE TROUGH -5 MO	-4 MO	-3 MO	-2 MO	-1 MO	0 MO
3	230.00	107.0	105.2	103.5	101.7	101.3	102.2	103.0	103.0	102.2	101.7	100.9	99.1
6	287.33	111.7	111.7	111.4	110.3	108.6	106.1	104.1	103.0	102.3	101.6	100.2	99.9
9	390.67	106.5	106.5	106.7	107.0	107.0	106.7	105.7	105.2	104.7	103.9	103.2	98.3
10	432.33	104.3	104.3	103.2	102.7	102.2	102.0	101.3	100.8	100.6	100.6	101.1	99.0
8	487.00	103.1	102.9	102.3	102.1	101.4	101.2	100.8	100.6	100.4	100.2	100.0	100.0
5	508.67	104.2	104.2	104.2	103.8	103.6	103.2	102.8	102.4	101.2	100.7	100.1	99.9
2	534.67	101.7	102.1	101.7	101.6	101.4	101.4	101.2	101.0	100.8	100.2	100.1	99.9
AL		738.5	736.9	732.9	729.2	725.5	722.9	719.0	716.2	712.3	709.0	705.5	696.1
RAGE		105.5	105.3	104.7	104.2	103.6	103.3	102.7	102.3	101.8	101.3	100.8	99.4
DEVIATION		2.5	2.2	2.5	2.6	2.4	1.8	1.4	1.3	1.1	1.0	0.8	0.5

OMIT THE FOLLOWING CYCLES

1945 9 1948 7 1949 10

		-11 MO	-10 MO	-9 MO	-8 MO	-7 MO	-6 MO	-5 MO	-4 MO	-3 MO	-2 MO	-1 MO	0 MO
AL		632.0	630.4	626.2	622.2	618.5	616.1	613.2	610.9	607.6	605.1	602.3	597.8
RAGE		105.3	105.1	104.4	103.7	103.1	102.7	102.2	101.8	101.3	100.8	100.4	99.6
DEVIATION		2.7	2.3	2.3	2.2	2.0	1.3	1.1	1.0	0.7	0.6	0.4	0.4

Output Table 4B-4S

EMPLOYEES IN NONAG ESTABLISHMENTS

HUNDRED THOUSAND PERSONS

8268

SPECIFIC ANALYSIS

DATE OF TROUGH	STANDING AT TROUGH	STANDINGS FIRST YEAR AFTER TROUGH +1 MO	+2 MO	+3 MO	+4 MO	+5 MO	+6 MO	+7 MO	+8 MO	+9 MO	+10 MO	+11 MO	+12 MO
1933 3	230.00	230.	233.	239.	245.	252.	256.	259.	259.	258.	259.	263.	26[illegible]
1938 6	287.33	287.	289.	292.	293.	297.	299.	298.	299.	302.	300.	302.	30[illegible]
1945 9	390.67	385.	388.	390.	397.	392.	402.	408.	413.	416.	420.	425.	42[illegible]
1949 10	432.33	432.	435.	435.	432.	439.	443.	446.	450.	454.	461.	463.	46[illegible]
1954 8	487.00	487.	488.	491.	493.	494.	496.	499.	501.	505.	507.	509.	50[illegible]
1958 5	508.67	509.	509.	512.	514.	514.	519.	520.	524.	526.	529.	532.	5[illegible]
1961 2	534.67	535.	535.	537.	539.	541.	542.	542.	543.	546.	547.	547.	5[illegible]

Output Table 4B-5S

DATE OF TROUGH	STANDING AT TROUGH	ABSOLUTE CHANGE FIRST YEAR AFTER TROUGH +1 MO	+2 MO	+3 MO	+4 MO	+5 MO	+6 MO	+7 MO	+8 MO	+9 MO	+10 MO	+11 MO	+12 MO
1933 3	230.00	0.0	3.0	9.0	15.0	22.0	26.0	29.0	29.0	28.0	29.0	33.0	37[illegible]
1938 6	287.33	-0.3	1.7	4.7	5.7	9.7	11.7	10.7	11.7	14.7	12.7	14.7	16[illegible]
1945 9	390.67	-5.7	-2.7	-0.7	6.3	1.3	11.3	17.3	22.3	25.3	29.3	34.3	37[illegible]
1949 10	432.33	-0.3	2.7	2.7	-0.3	6.7	10.7	13.7	17.7	21.7	28.7	30.7	32[illegible]
1954 8	487.00	0.0	1.0	4.0	6.0	7.0	9.0	12.0	14.0	18.0	20.0	22.0	22[illegible]
1958 5	508.67	0.3	0.3	3.3	5.3	5.3	10.3	11.3	15.3	17.3	20.3	23.3	26[illegible]
1961 2	534.67	0.3	0.3	2.3	4.3	6.3	7.3	7.3	8.3	11.3	12.3	12.3	15[illegible]
TOTAL		-5.7	6.3	25.3	42.3	58.3	86.3	101.3	118.3	136.3	152.3	170.3	187[illegible]
AVERAGE		-0.8	0.9	3.6	6.0	8.3	12.3	14.5	16.9	19.5	21.8	24.3	26[illegible]
AVE DEVIATION		1.4	1.3	1.9	2.6	4.3	3.9	5.0	5.2	4.7	6.2	7.1	7[illegible]

OMIT THE FOLLOWING CYCLES

1945 9 1948 7 1949 10

	+1 MO	+2 MO	+3 MO	+4 MO	+5 MO	+6 MO	+7 MO	+8 MO	+9 MO	+10 MO	+11 MO	+12 MO
TOTAL	0.0	9.0	26.0	36.0	57.0	75.0	84.0	96.0	111.0	123.0	136.0	15[illegible]
AVERAGE	0.0	1.5	4.3	6.0	9.5	12.5	14.0	16.0	18.5	20.5	22.7	2[illegible]
AVE DEVIATION	0.2	0.9	1.7	3.0	4.2	4.5	5.0	4.9	4.2	5.6	6.3	[illegible]

Output Table 4B-6S

DATE OF TROUGH	STANDING AT TROUGH	RELATIVE STANDING FIRST YEAR AFTER TROUGH +1 MO	+2 MO	+3 MO	+4 MO	+5 MO	+6 MO	+7 MO	+8 MO	+9 MO	+10 MO	+11 MO	+12 MO
1933 3	230.00	100.0	101.3	103.9	106.5	109.6	111.3	112.6	112.6	112.2	112.6	114.3	11[illegible]
1938 6	287.33	99.9	100.6	101.6	102.0	103.4	104.1	103.7	104.1	105.1	104.4	105.1	10[illegible]
1945 9	390.67	98.5	99.3	99.8	101.6	100.3	102.9	104.4	105.7	106.5	107.5	108.8	10[illegible]
1949 10	432.33	99.9	100.6	100.6	99.9	101.5	102.5	103.2	104.1	105.0	106.6	107.1	10[illegible]
1954 8	487.00	100.0	100.2	100.8	101.2	101.4	101.8	102.5	102.9	103.7	104.1	104.5	10[illegible]
1958 5	508.67	100.1	100.1	100.7	101.0	101.0	102.0	102.2	103.0	103.4	104.0	104.6	10[illegible]
1961 2	534.67	100.1	100.1	100.4	100.8	101.2	101.4	101.4	101.6	102.1	102.3	102.3	10[illegible]
TOTAL		698.5	702.2	707.9	713.1	718.5	726.0	730.0	733.9	738.0	741.6	746.7	75[illegible]
AVERAGE		99.8	100.3	101.1	101.9	102.6	103.7	104.3	104.8	105.4	105.9	106.7	10[illegible]
AVE DEVIATION		0.4	0.5	0.9	1.4	2.2	2.3	2.4	2.5	2.2	2.6	2.9	

OMIT THE FOLLOWING CYCLES

1945 9 1948 7 1949 10

	+1 MO	+2 MO	+3 MO	+4 MO	+5 MO	+6 MO	+7 MO	+8 MO	+9 MO	+10 MO	+11 MO	+12 MO
TOTAL	599.9	602.8	608.1	611.5	618.1	623.1	625.5	628.2	631.5	634.1	638.0	64[illegible]
AVERAGE	100.0	100.5	101.3	101.9	103.0	103.8	104.3	104.7	105.3	105.7	106.3	10[illegible]
AVE DEVIATION	0.1	0.4	0.9	1.6	2.3	2.6	2.8	2.6	2.3	2.6	2.9	

Output Table 4B-4S

MPLOYEES IN NONAG ESTABLISHMENTS

HUNDRED THOUSAND PERSONS 8268

PECIFIC ANALYSIS

ATE OF OUGH	STANDING AT PEAK	-11 MO	-10 MO	STANDINGS STANDING -9 MO	-8 MO	-7 MO	-6 MO	ONE YEAR BEFORE TROUGH -5 MO	-4 MO	-3 MO	-2 MO	-1 MO	0 MO
33 3	332.00	246.	242.	238.	234.	233.	235.	237.	237.	235.	234.	232.	228.
38 6	320.67	321.	321.	320.	317.	312.	305.	299.	296.	294.	292.	288.	287.
45 9	426.33	416.	416.	417.	418.	418.	417.	413.	411.	409.	406.	403.	384.
49 10	450.00	451.	451.	446.	444.	442.	441.	438.	436.	435.	435.	437.	428.
54 8	503.67	502.	501.	498.	497.	494.	493.	491.	490.	489.	488.	487.	487.
58 5	531.00	530.	530.	530.	528.	527.	525.	523.	521.	515.	512.	509.	508.
61 2	544.66	544.	546.	544.	543.	542.	542.	541.	540.	539.	536.	535.	534.

Output Table 4B-7S

TE OF UGH	STANDING AT PEAK	-11 MO	-10 MO	ABSOLUTE CHANGE -9 MO	-8 MO	-7 MO	-6 MO	ONE YEAR BEFORE TROUGH -5 MO	-4 MO	-3 MO	-2 MO	-1 MO	0 MO
3 3	332.00	-86.0	-90.0	-94.0	-98.0	-99.0	-97.0	-95.0	-95.0	-97.0	-98.0	-100.0	-104.0
8 6	320.67	0.3	0.3	-0.7	-3.7	-8.7	-15.7	-21.7	-24.7	-26.7	-28.7	-32.7	-33.7
5 9	426.33	-10.3	-10.3	-9.3	-8.3	-8.3	-9.3	-13.3	-15.3	-17.3	-20.3	-23.3	-42.3
9 10	450.00	1.0	1.0	-4.0	-6.0	-8.0	-9.0	-12.0	-14.0	-15.0	-15.0	-13.0	-22.0
4 8	503.67	-1.7	-2.7	-5.7	-6.7	-9.7	-10.7	-12.7	-13.7	-14.7	-15.7	-16.7	-16.7
8 5	531.00	-1.0	-1.0	-1.0	-3.0	-4.0	-6.0	-8.0	-10.0	-16.0	-19.0	-22.0	-23.0
1 2	544.66	-0.7	1.3	-0.7	-1.7	-2.7	-2.7	-3.7	-4.7	-5.7	-8.7	-9.7	-10.7
TAL		-98.3	-101.3	-115.3	-127.3	-140.3	-150.3	-166.3	-177.3	-192.3	-205.3	-217.3	-252.3
ERAGE		-14.0	-14.5	-16.5	-18.2	-20.0	-21.5	-23.8	-25.3	-27.5	-29.3	-31.0	-36.0
E DEVIATION		20.6	21.6	22.1	22.8	22.6	21.6	20.4	19.9	19.9	19.6	20.2	21.2

OMIT THE FOLLOWING CYCLES

1945 9 1948 7 1949 10

		-11 MO	-10 MO	-9 MO	-8 MO	-7 MO	-6 MO	-5 MO	-4 MO	-3 MO	-2 MO	-1 MO	0 MO
AL		-88.0	-91.0	-106.0	-119.0	-132.0	-141.0	-153.0	-162.0	-175.0	-185.0	-194.0	-210.0
RAGE		-14.7	-15.2	-17.7	-19.8	-22.0	-23.5	-25.5	-27.0	-29.2	-30.8	-32.3	-35.0
DEVIATION		23.8	24.9	25.4	26.1	25.7	24.5	23.2	22.7	22.6	22.4	22.7	23.0

Output Table 4B-8S

E OF GH	STANDING AT PEAK	-11 MO	-10 MO	RELATIVE STANDING -9 MO	-8 MO	-7 MO	-6 MO	ONE YEAR BEFORE TROUGH -5 MO	-4 MO	-3 MO	-2 MO	-1 MO	0 MO
3	332.00	74.1	72.9	71.7	70.5	70.2	70.8	71.4	71.4	70.8	70.5	69.9	68.7
6	320.67	100.1	100.1	99.8	98.9	97.3	95.1	93.2	92.3	91.7	91.1	89.8	89.5
9	426.33	97.6	97.6	97.8	98.0	98.0	97.8	96.9	96.4	95.9	95.2	94.5	90.1
10	450.00	100.2	100.2	99.1	98.7	98.2	98.0	97.3	96.9	96.7	96.7	97.1	95.1
8	503.67	99.7	99.5	98.9	98.7	98.1	97.9	97.5	97.3	97.1	96.9	96.7	96.7
5	531.00	99.8	99.8	99.8	99.4	99.2	98.9	98.5	98.1	97.0	96.4	95.9	95.7
2	544.66	99.9	100.2	99.9	99.7	99.5	99.5	99.3	99.1	99.0	98.4	98.2	98.0
AL		671.4	670.3	667.0	663.9	660.6	658.0	654.1	651.5	648.1	645.2	642.1	633.8
RAGE		95.9	95.8	95.3	94.8	94.4	94.0	93.4	93.1	92.6	92.2	91.7	90.5
DEVIATION		6.2	6.5	6.7	7.0	6.9	6.6	6.4	6.4	6.5	6.5	6.8	6.7

OMIT THE FOLLOWING CYCLES

1945 9 1948 7 1949 10

		-11 MO	-10 MO	-9 MO	-8 MO	-7 MO	-6 MO	-5 MO	-4 MO	-3 MO	-2 MO	-1 MO	0 MO
L		573.8	572.7	569.2	565.8	562.5	560.2	557.3	555.1	552.2	549.9	547.6	543.7
AGE		95.6	95.5	94.9	94.3	93.8	93.4	92.9	92.5	92.0	91.7	91.3	90.6
DEVIATION		7.2	7.5	7.7	7.9	7.9	7.5	7.2	7.1	7.2	7.3	7.6	7.7

Output Table 4B-4S

EMPLOYEES IN NONAG ESTABLISHMENTS

HUNDRED THOUSAND PERSONS — 8268

SPECIFIC ANALYSIS

STANDINGS

DATE OF TROUGH	STANDING AT PEAK	+1 MO	+2 MO	STANDING +3 MO	+4 MO	+5 MO	+6 MO	FIRST YEAR AFTER TROUGH +7 MO	+8 MO	+9 MO	+10 MO	+11 MO	+12 MO
1933 3	332.00	230.	233.	239.	245.	252.	256.	259.	259.	258.	259.	263.	26
1938 6	320.67	287.	289.	292.	293.	297.	299.	298.	299.	302.	300.	302.	30
1945 9	426.33	385.	388.	390.	397.	392.	402.	408.	413.	416.	420.	425.	42
1949 10	450.00	432.	435.	435.	432.	439.	443.	446.	450.	454.	461.	463.	46
1954 8	503.67	487.	488.	491.	493.	494.	496.	499.	501.	505.	507.	509.	5
1958 5	531.00	509.	509.	512.	514.	514.	519.	520.	524.	526.	529.	532.	5
1961 2	544.66	535.	535.	537.	539.	541.	542.	542.	543.	546.	547.	547.	5

Output Table 4B-7S

ABSOLUTE CHANGE

DATE OF TROUGH	STANDING AT PEAK	+1 MO	+2 MO	ABSOLUTE CHANGE +3 MO	+4 MO	+5 MO	+6 MO	FIRST YEAR AFTER TROUGH +7 MO	+8 MO	+9 MO	+10 MO	+11 MO	+12
1933 3	332.00	-102.0	-99.0	-93.0	-87.0	-80.0	-76.0	-73.0	-73.0	-74.0	-73.0	-69.0	-6
1938 6	320.67	-33.7	-31.7	-28.7	-27.7	-23.7	-21.7	-22.7	-21.7	-18.7	-20.7	-18.7	-1
1945 9	426.33	-41.3	-38.3	-36.3	-29.3	-34.3	-24.3	-18.3	-13.3	-10.3	-6.3	-1.3	[illegible]
1949 10	450.00	-18.0	-15.0	-15.0	-18.0	-11.0	-7.0	-4.0	0.0	4.0	11.0	13.0	1
1954 8	503.67	-16.7	-15.7	-12.7	-10.7	-9.7	-7.7	-4.7	-2.7	1.3	3.3	5.3	[illegible]
1958 5	531.00	-22.0	-22.0	-19.0	-17.0	-17.0	-12.0	-11.0	-7.0	-5.0	-2.0	1.0	[illegible]
1961 2	544.66	-9.7	-9.7	-7.7	-5.7	-3.7	-2.7	-2.7	-1.7	1.3	2.3	2.3	[illegible]
TOTAL		-243.3	-231.3	-212.3	-195.3	-179.3	-151.3	-136.3	-119.3	-101.3	-85.3	-67.3	-5
AVERAGE		-34.8	-33.0	-30.3	-27.9	-25.6	-21.6	-19.5	-17.0	-14.5	-12.2	-9.6	-
AVE DEVIATION		21.1	20.4	19.6	17.3	18.0	16.3	16.2	17.3	18.2	19.8	19.6	1

OMIT THE FOLLOWING CYCLES

1945 9 1948 7 1949 10

	+1 MO	+2 MO	+3 MO	+4 MO	+5 MO	+6 MO	+7 MO	+8 MO	+9 MO	+10 MO	+11 MO	+12
TOTAL	-202.0	-193.0	-176.0	-166.0	-145.0	-127.0	-118.0	-106.0	-91.0	-79.0	-66.0	-5
AVERAGE	-33.7	-32.2	-29.3	-27.7	-24.2	-21.2	-19.7	-17.7	-15.2	-13.2	-11.0	-
AVE DEVIATION	22.8	22.3	21.2	19.8	18.6	18.4	18.8	19.8	20.8	22.4	21.9	[illegible]

Output Table 4B-8S

RELATIVE STANDING

DATE OF TROUGH	STANDING AT PEAK	+1 MO	+2 MO	RELATIVE STANDING +3 MO	+4 MO	+5 MO	+6 MO	FIRST YEAR AFTER TROUGH +7 MO	+8 MO	+9 MO	+10 MO	+11 MO	+12
1933 3	332.00	69.3	70.2	72.0	73.8	75.9	77.1	78.0	78.0	77.7	78.0	79.2	[illegible]
1938 6	320.67	89.5	90.1	91.1	91.4	92.6	93.2	92.9	93.2	94.2	93.6	94.2	[illegible]
1945 9	426.33	90.3	91.0	91.5	93.1	91.9	94.3	95.7	96.9	97.6	98.5	99.7	10
1949 10	450.00	96.0	96.7	96.7	96.0	97.6	98.4	99.1	100.0	100.9	102.4	102.9	10
1954 8	503.67	96.7	96.9	97.5	97.9	98.1	98.5	99.1	99.5	100.3	100.7	101.1	1
1958 5	531.00	95.9	95.9	96.4	96.8	96.8	97.7	97.9	98.7	99.1	99.6	100.2	1
1961 2	544.66	98.2	98.2	98.6	99.0	99.3	99.5	99.5	99.7	100.2	100.4	100.4	1
TOTAL		635.9	639.0	643.7	647.9	652.2	658.8	662.3	666.0	669.9	673.2	677.6	6
AVERAGE		90.8	91.3	92.0	92.6	93.2	94.1	94.6	95.1	95.7	96.2	96.8	
AVE DEVIATION		6.7	6.4	6.1	5.7	5.4	5.1	5.2	5.4	5.6	5.9	5.8	

OMIT THE FOLLOWING CYCLES

1945 9 1948 7 1949 10

	+1 MO	+2 MO	+3 MO	+4 MO	+5 MO	+6 MO	+7 MO	+8 MO	+9 MO	+10 MO	+11 MO	+12
TOTAL	545.6	547.9	552.2	554.8	560.3	564.5	566.6	569.1	572.3	574.7	578.0	5
AVERAGE	90.9	91.3	92.0	92.5	93.4	94.1	94.4	94.9	95.4	95.8	96.3	
AVE DEVIATION	7.7	7.4	7.0	6.6	6.1	5.9	6.0	6.1	6.3	6.7	6.4